# AN INTRODUCTION
# TO SEMICONDUCTOR
# MICROTECHNOLOGY

# AN INTRODUCTION TO SEMICONDUCTOR MICROTECHNOLOGY

## (Second Edition)

### D. V. Morgan

Professor of Microelectronics
School of Electrical, Electronic & Systems Engineering
University of Wales College of Cardiff, Wales

### K. Board

Professor of Electronic Engineering
Department of Electrical and Electronic Engineering
University of Wales, Swansea, Wales

**JOHN WILEY & SONS**

Chichester • New York • Brisbane • Toronoto • Singapore

*Other Wiley Editorial Offices*

John Wiley & Sons, Inc., 605 Third Avenue,
New York, NY 10158-0012, USA

Jacaranda Wiley Ltd, G.P.O. Box 859, Brisbane,
Queensland 4001, Australia

John Wiley & Sons (Canada) Ltd, 22 Worcester Road,
Rexdale, Ontario M9W 1L1, Canada

John Wiley & Sons (SEA) Pte Ltd, 37 Jalan Pemimpin #05-04,
Block B, Union Industrial Building, Singapore 2057

**Library of Congress Cataloging-in-Publication Data:**
Morgan, D. V.
    An introduction to semiconductor microtechnology / D. V. Morgan, K.
Board. – 2nd ed.
      p.  cm.
    Includes bibliographical references.
    ISBN 0 471 92478 4
    1. Semiconductors—Design and construction.    I. Board, K.
II. Title.
TK7871.85.M585   1990
621.381′52—dc20                        89–70572

**British Library Cataloguing in Publication Data:**
Morgan, D. V. (David Vernon), *1941–*
    Introduction to semiconductor microtechnology. – 2nd ed.
    1. Semiconductor devices
    I. Title    II. Board, K. (Kenneth)
    621.3815′2

ISBN 0 471 92478 4

Typeset by APS, Salisbury & Petersfield
Printed and bound in Great Britain by
Biddles Ltd, Guildford and King's Lynn

We dedicate this book to our families

*Jean, Suzanne* and *Dyfrig Morgan*

*Meriel, Meirion* and *Alun Board*

# Contents

# Preface

There are few people in the industrialized nations who have not been influenced by the staggering developments which have taken place in semiconductor microtechnology during the past two decades. For the majority of people this has taken the form of very fundamental influences in both working and leisure activities. Terms such as robotics, word processing, electronic games are now in everyday usage. To the professional electronic engineer these innovations have heralded a major revolution in practice with drastic reductions in cost, size, and power consumption. With this background in mind this book is written primarily as a 'first exposure'—an introductory text to the topic of semiconductor microtechnology. The minimum of mathematics is intended to make the text suitable to a wide range of students who wish to gain some understanding of the basic principles upon which this very important technology is based. Chapter 1 presents an introduction to the development of microtechnology, introducing some of the basic concepts invoked. Chapter 2 is concerned with the growth and preparation of semiconducting materials. In Chapter 3 the processes of doping and type-conversion are presented, together with a discussion of the twin technologies of diffusion and ion implantation. Chapter 4 deals with oxidation, Chapter 5 the process of photolithography, and Chapter 6 metallization, interconnections, and bonding. In Chapters 7 and 8, the structure and processing of circuit components (Chapter 7) and integrated circuits are considered. The concluding chapter, Chapter 9, speculates on future trends in semiconductor microtechnology.

We wish to express our gratitude to all the authors, publishers, and companies who have allowed us to modify and reproduce published data, to Dr C. E. C. Wood and Professor L. F. Eastman (Cornell University) for many stimulating discussions on the topic of semiconductor technology. Finally we are indebted to our respective families for their patience during the preparation of this manuscript.

D. V. MORGAN    *University of Wales, Cardiff*
K. BOARD    *University of Wales, Swansea*    November 1988

# List of Symbols

| | |
|---|---|
| $B$ | Magnetic field strength |
| $C$ | Capacitance |
| $C_f$ | Correction factor for 4-point probe |
| $c$ | Velocity of light |
| $D$ | Diffusion coefficient of impurity |
| $D_0$ | Diffusion coefficent of oxidant in $SiO_2$, Diffusion coefficient at $T \to \infty$ |
| $d$ | Junction depth |
| $E_a$ | Diffusion activation energy |
| $E_g$ | Band gap energy |
| $E_s$ | Formation energy of a vacancy |
| $E_1$ | Activation energy of oxidant diffusing in $SiO_2$ |
| $E_2$ | Activation energy of reaction to form $SiO_2$ |
| $F$ | Electric field strength |
| $F_f$ | Oxidant flux |
| $F_i$ | Oxidant flux at $SiO_2$ interface |
| $F_0$ | Oxidant flux through $SiO_2$ |
| $f$ | Resultant force on charge carrier |
| $I, I_x, I_y, I_z$ | Current and its cartesian components |
| $K_1, K_2$ | Constants relating to the thermal oxidation process |
| $k$ | Boltzmann's constant |
| $k_i$ | Reaction rate constant of thermal oxidation process |
| $m^*$ | Effective mass of charge carrier ($m_e^*$ electrons, $m_h^*$ holes) |
| $N$ | Oxidant concentration, Impurity concentration |
| $N_A$ | Acceptor concentration |
| $N_D$ | Donor concentration |
| $N_i$ | Oxidant concentration at interface |
| $N_m$ | $SiO_2$ molecules per unit volume |
| $N_s$ | Oxidant concentration at oxide surface |
| $N_T$ | Total impurity concentration |
| $N_0$ | Surface atomic concentration |
| $n$ | Free electron concentration |
| $N_1, N_2$ | Volume concentration of atoms |
| $n_1, n_2$ | Atomic density in atomic planes 1 and 2 respectively |

| | |
|---|---|
| $n_i$ | Intrinsic carrier concentration |
| $p$ | Free hole concentration |
| $Q$ | Area density of atoms diffusing into solid |
| $q$ | Electronic charge |
| $R$ | Resistance |
| $R_H$ | Hall coefficient |
| $R_p$ | Mean projected range of implanted ion |
| $R_\perp$ | Mean transverse range of implanted ion |
| $S$ | Device area |
| $T$ | Absolute temperature |
| $T_f$ | Freeze-out temperature |
| $t$ | Time |
| $u$ | Mean drift velocity |
| $V$ | Potential |
| $V_T$ | Threshold voltage (enhancement MOST) |
| $v, v_x$ | Particle velocity |
| $x_i$ | Initial oxide thickness |
| $x_0$ | Oxide thickness |
| $\varepsilon_s$ | Permittivity of silicon |
| $\phi$ | Net flux of diffusing atoms |
| $\lambda$ | Mean free path between collisions |
| $\mu$ | Mobility |
| $\mu_H$ | Hall mobility |
| $\mu_M$ | GMR mobility |
| $\nu$ | Atomic vibrational frequency |
| $\nu_j$ | Atomic jump frequency |
| $\sigma_p$ | Range straggling |
| $\rho$ | Resistivity |
| $\rho_\square$ | Sheet resistivity |
| $\sigma$ | Conductivity |
| $\tau$ | Mean time between collisions |
| $\tau_0$ | Initial oxidation time |
| $\chi$ | Electron affinity |

# CHAPTER 1

# The Development of Semiconductor Technology

## Instructional Objectives

*The first chapter acquaints the student with the basic reasons for producing integrated circuits (ICs). It assumes the student already knows, in general, the definition of an IC. After reading this chapter, you should be able to:*

a. Explain the cost advantage of integrated circuits over discrete circuits.
b. Explain some of the problems associated with ICs, such as the yield.
c. Explain a simple batch processing technique for making ICs.
d. Explain the relationship between yield and chip size.

## Self-evaluation Questions

*Watch for the answers to these questions as you read the chapter. They will help point out the important ideas presented.*

a. What is the typical diameter of silicon wafers used to make ICs and how has this changed over the past 15 years?
b. What are the cost advantages of ICs over discrete circuits?
c. What are the factors governing chip size on a wafer?

## 1.1 INTRODUCTION

With the invention of a solid state amplifying device in 1947 by Shockley, Bardeen, and Brattain the possibility of 'integrating' not one, but many transistors within the same crystal has existed. Hitherto the valve or vacuum tube provided the only means of amplifying an electrical signal and integration in this context was never feasible.

The economic advantages and improvements in performance that occur in integrating many devices within the same crystal or chip are, as we discuss later in this chapter, very great indeed. There has thus grown up over the last twenty years a technology which has enabled increasing numbers of devices to be placed on a single chip.

Sophisticated new techniques have been developed and new ones are emerging in an effort to continue this trend towards higher integration levels. This has reached the point where a new discipline, that of 'semiconductor microtechnology' has emerged, and this constitutes the main purpose of this book.

The development of this technology has been swift but expensive. Since the first 'integrated circuit' in 1959 the number of components per chip has almost doubled every year, so that it is now well in excess of 100 000 (refer to Fig. 9,3(a). The pressures to bring about such a revolution in electronics have to be powerful indeed and most of this chapter will be devoted to a discussion of the reasons for it. In the final part of the chapter a simple prototype integrated circuit is discussed in detail. Although the circuit itself is simple, its fabrication involves most of the processing steps found in more complicated integrated circuits and it will therefore serve to illustrate the individual processes and how they are linked together to produce an overall sequence.

## 1.2  ADVANTAGES OF INTEGRATION

An interesting analogy may be drawn between the development of the integrated circuit and that of the automobile. Society has been changed profoundly by both over a relatively short period, and the reasons in both cases were primarily economic.

It was not the invention of the petrol engine in 1884 that initiated the large-scale use of the automobile in society; for many years it was available to only a small number of very wealthy people. Rather, it was the invention in 1903 of mass-production techniques by Henry Ford that gave rise to its widespread use.

Over the last few years the impact of electronics on all aspects of our society has been rapid and dramatic, although the basic electronic functions and operations in use were known for many years before this. The reason is that a technique known as 'batch processing' has made it far cheaper than before. The technique has similarities with mass production and has had similar consequences. The dramatic reduction in the cost of electronic functions has, in parallel with the automobile, made it available to people and applications that would have been ruled out for economic reasons.

### 1.2.1  Batch processing

It is readily possible to make even complex circuits on areas of silicon less than 8 mm square. Typically, silicon wafers of 100–150 mm diameter are used so that each contains about 120–300 circuits. Each wafer is passed through the various processing stages so that each circuit is treated simultaneously or as a batch. Furthermore, it is usual for the manufacturer to process, say, several hundred wafers together (i.e. up to 60 000 circuits). Take, for example, one of these processing stages known as diffusion (dealt with in detail in Chapter 3).

If it costs say £5000 to run the diffusion equipment, pay the salaries for that run, and two hundred 150 mm wafers are processed together, the cost per circuit is only £5000 $\div$ (200 $\times$ 300), about 8 p. (Actually one should consider here only the number of *working* circuits since some will malfunction; this problem of 'yield' is discussed later in the chapter.) Thus, although the actual cost is high, the added cost per circuit is very low.

There are, of course, stages in the manufacture of the circuit which have to be carried out on each circuit individually: these include packaging and testing. The costs are not divided by the number produced but have to be added directly to the unit cost. Thus they are usually the major components in the cost of present-day integrated circuits.

## 1.3  LEVELS OF INTEGRATION

The more components a manufacturer is able to place on a single semiconductor chip, the higher is the level of integration he achieves. At one extreme he may use single transistors, each separately packaged, and separate resistors, capacitors, and other components. This is the fully *discrete* approach. At the other extreme he may place the total electronic system on the chip and here he achieves the highest possible level of integration. There is a whole range of levels between these limiting cases, and the decision on which is optimum is a complex economic one depending on such factors as availability of integrated circuits and projected market size.

In order to illustrate the powerful cost advantages of integration we consider the example of an electronic system manufacturer who wishes to manufacture a system containing 10 000 components. His marketing manager advises him that $10^6$ of these may be sold over a five-year period, after which the product is likely to be superseded.

The total design costs for the system are £40 000. Every 50 components are connected together on a printed-circuit board and we assume that the latter cost £5 to produce. Table 1.1 illustrates three different approaches; one fully discrete, one fully integrated, and one intermediate case. The design costs are negligible since they are disposed over $10^6$ units. The individual 'component' cost (here defined as a single packaged silicon chip) increases only slightly with the number

Table 1.1    Cost comparison of the three different levels of integration

| Unit cost | Discrete approach | Partly integrated approach | Fully integrated |
|---|---|---|---|
| Design costs | £40 000/10$^6$ = £0.04 | £40 000/10$^6$ = £0.04 | £40 000/10$^6$ = £0.04 |
| Component costs | 10 000 @£0.5 £5000 | 100 ICs (100 comp/chip) @ £3 = £300 | 1 IC @ £10 = £10 |
| P.C. board | 200 boards @ £5 = £1000 | 2 P.C. boards @ £5 = £10 | — |
| Assembly and packaging | £50 | £20 | £20 |
| Total unit cost | £6050 | £330 | £30 |

of components on it; the major differences in the unit cost are because of the differences in the numbers of components and p.c. boards required for each implementation.

There are thus seen to be very powerful cost incentives to move to higher levels of integration. In addition there is a size and weight advantage that is clear from Table 1.1. Two hundred p.c. boards with all their interconnections would fill a large cabinet, whereas the 2 p.c. board intermediate version would occupy only small bench top instrument casing. The final version could, depending on the peripherals required, fit in a hand-held case.

Although each of the three versions performs the same electronic function there would possibly be some performance advantage in the fully integrated form because of the elimination of long lead lengths. However a much more important advantage is apparent in the latter implementation — reliability. The most common causes of failure tend to be at the interconnections between the components and the rest of the circuit. Most of these have been placed on the chip in the fully integrated version and therefore eliminated, making this by far the most reliable, as well as the cheapest and smallest, of the three versions considered.

In the rather oversimplified example just discussed, a number of factors which modify this simple picture should be mentioned.

## 1.4   THE PROBLEM OF LARGE VOLUMES

In the previous example we discussed the situation from the point of view of the electronic system manufacturer who purchased his discrete or integrated circuits from an IC manufacturer. As far as the latter is concerned, unless he can fill his production lines close to full capacity, it is not worth his while to produce them.

If, for example, he has a 75 mm line with 10 production runs per week, with a circuit size of 5 × 5 mm, he would produce 1.8 million annually (assuming only 10% of his output functioned correctly at the finish). The system manufacturer in our example only requires $2 \times 10^5$ annually for his product so that unless there are other customers the IC manufacturer may not consider it worthwhile to produce the fully integrated version. The system manufacturer may then be forced to move to lower levels of integration.

There is a general rule that the more complex any system becomes, the more specialized it is and hence the smaller is the number of applications. Thus a move to higher complexity means that fewer are likely to be sold, which may mean in turn that they are not made because it is not economic.

The advent of the microprocessor was a specific attempt to circumvent this problem. It combines a high degree of complexity with 'programmability' so that it may be used to perform a large number of different tasks. It has made it feasible for IC manufacturers to achieve high volume sales for very complex electronic systems, and made available to the systems and circuit designers a powerful new tool at very low cost.

### 1.4.1 Yield

The number of 'working' circuits at the end of the process sequence is of particular concern to the integrated circuit manufacturer. If the process is new and innovative, then it is likely that in the early stages the yields will be low. The greater the number of stages in the process the lower will be the yield, since each stage carries with it a finite probability that a malfunction will be induced in some of the circuits. For the case where the fraction $f$ still working after each stage is the same for each stage, the fraction working after $N$ stages is

$$Y = f^N. \tag{1.1}$$

For example, for a 30-stage process, with 10% lost at each stage ($f = 0.9$), the overall yield is only 4%. If the yield per stage $f$ is 0.98, the overall yield increases to 54%. However, if the number of process steps, $N$, is incresed to 50, then the yield drops to 36% with the same yield per stage.

### 1.4.2 Maximum complexity

Another factor that affects the decision on which integration level is appropiate is that there is at a given moment in time a maximum complexity that the IC manufacturer can achieve with adequate yields. At one end of the scale he has a minimum feature size which sets a limit on the component density he can achieve. If he attempts to make a circuit component smaller than that which this minimum feature size will allow, he will suffer a dramatic decrease in yield because the misalignment between successive layers exceeds what is necessary to produce a working circuit. Thus, for example, a connection between two components may not be made, leaving an open circuit.

At the other end of the scale there is maximum chip size above which the yield again decreases rapidly. The starting semiconductor substrate or wafer has a fixed number of defects per unit area as illustrated in Fig. 1.1. If we assume that each defect is effective in preventing a chip which incorporates it from working, then the maximum yield that the manufacturer can achieve is zero with 4 chips per wafer, 75% with 16 and 98% with 256. It would be inadvisable for him to operate any process at a chip size corresponding of 4 per wafer since his yields are severely limited before he starts. Of course, in practice such defects would be spread randomly over the surface so that statistically these percentages would vary according to where on the wafer the defects occurred.

Both factors considered together mean that there is a maximum achievable

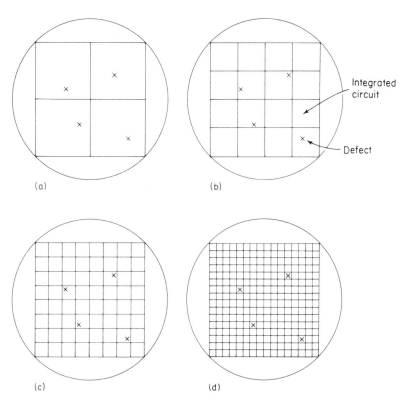

Fig. 1.1  Relationship between yield and chip size: (a) yield = 0; (b) yield = 75%; (c) yield = 94%; (d) yield = 98%

complexity in terms of the number of components per chip. This can only be increased by

(a) The supplier of the substrates producing lower defect densities in his starting material.
(b) Obtaining new processing equipment which permits a smaller minimum feature size.

Both are outside the IC manufacturer's control and are improvements that only come slowly with time and at considerable expense.

Subsequent chapters in this book are concerned with specific stages in the fabrication of integrated circuits. In order to give first an overall view of the complete technology the rest of this chapter will be devoted to a discussion of a simple prototype IC which, although in circuit terms is unrealistic as a commercial IC, nevertheless contains all the individual process steps used in much more complex circuits.

## 1.5 BATCH PROCESSING OF A PROTOTYPE INTEGRATED CIRCUIT

A cross section and plan view of the circuit, in this case a simple pn junction diode, is shown in Fig. 1.2. The connection pattern, consisting of a thin layer of aluminium, is insulated from most of the chip by a layer of silicon dioxide. Only where the actual pn junction is formed is the silicon dioxide not present, so that (a) p-type impurities can be introduced or diffused into the n-type substrate to form the junction and (b) the aluminium can make electrical contact with the p-region.

Thus two pattern-formation steps can be identified. In the first a rectangular 'hole' has to be formed in the $SiO_2$ layer. When the p-type impurities are then introduced, the insulating layer prevents them from entering the semicondutor in all places except where the $SiO_2$ layer has been removed. Thus the pn junction is formed only in the localized region where required. This is illustrated in Fig. 1.3 at the first level. Secondly, a pattern has to be formed in the subsequent aluminium layer in order to define the shape of the electrical connection to the p-region and to isolate it from other parts of the circuit, as would of course be necessary if there were other circuit elements, such as resistors, capacitors, or transistors present. This pattern is illustrated in the second level shown in Fig. 1.3. The second electrical contact, that to the n-region, is made to the bottom face of the substrate and is often formed when the chip is mounted in its package. However, it is important to note that no pattern formation stage is necessary in this case. When each of these pattern-forming stages is carried out, they are carried out simultaneously by using an array of identical patterns, shown for each level in Fig. 1.3.

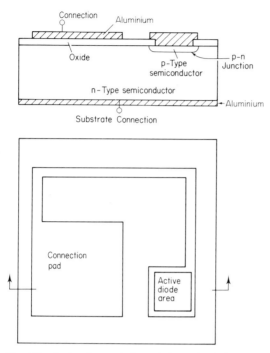

Fig. 1.2   A simple prototype integrated circuit to illustrate the various processing stages

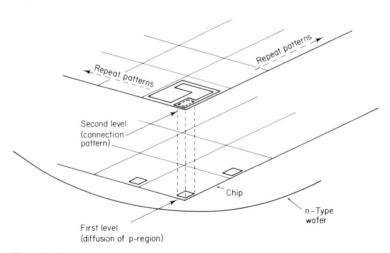

Fig. 1.3   Batch processing of pn junction diode illustrating the two patterning levels necessary

We can now summarize the processing of our simple circuit as follows (The chapter where each process is discussed is indicated):

1. Start with 1 'slice' or 'wafer' (Chapter 2) of n-type silicon. (This is typically 100–150 mm diameter and 0.5 mm thick).

2. Cover surface with an insulating layer of $SiO_2$ ('Oxidation', Chapter 4).

level    3. Open holes in $SiO_2$ for formation of p-region ('Photolithography',
1          Chapter 5).

4. Introduce p-region impurities ('Diffusion', Chapter 3).

5. Cover surface with aluminium ('Metallization', Chapter 6).

level    6. Define electrode pattern in aluminium ('Photolithography').
2        7. Cover back surface with suitable alloy (e.g. gold-antimony).

8. Carry out heat treatment to form low-resistance contacts between aluminium contact and the p-region, and between the back contact and the n-region.

Each of the processes 1–7 is performed on all circuits together on the wafer. The slice at this stage is sawn or scribed and diced into chips, each containing one diode with its connections. All subsequent processing, such as mounting in its package and wire-bonding will be carried out on each chip. This transition from batch to individual processing has an influence on the cost of the final circuit and has been discussed earlier in this chapter. The dicing, mounting, wire bonding, and hermetic sealing of our final circuit are shown in Fig. 1.4. A diamond saw is used to cut the silicon along tracks, provided for at the mask stages. The dice are mounted on headers using some combination of alloying and ultrasonic vibration and the top connection made with fine gold or aluminium wire between the contact on the chip and a pin connection. Finally a cover is put on which provides a hermetic seal so that the circuit is protected from changes in atmospheric conditions and physical damage. A detailed discussion of these techniques is given in Chapter 6. Figure 1.5 shows photographs of (a) A typical low-pin count package and (b) an IC chip mounted in a Dual-In-Line (DIL) 32pin package.

Following this overview of integrated circuit fabrication at a very simple level a detailed discussion of each of the process steps will be undertaken in subsequent chapters. Figure 1.6 shows a typical clean-room facility.

## Problems

1.1 The number of working circuits at the end of a process sequence is of particular concern to the IC manufacturer. Given that a manufacturer loses 5 percent of a lot at each stage and there are nine stages, determine the overall yield.

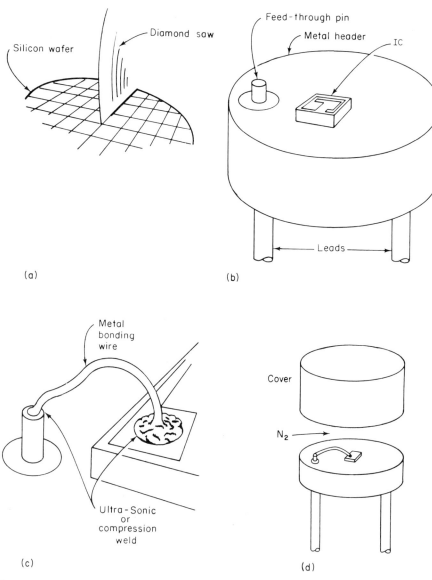

Fig. 1.4   (a) Sawing, (b) mounting, (c) wire bonding, and (d) hermetic sealing of prototype integrated circuit

(a)

Fig. 1.5   (a) Photograph of a single device (transistor) mounted on a T05 header (courtesy Airpax Corporation, (USA);

(b)

Fig. 1.5 (*continued*)    (b) corresponding photograph of a full silicon IC chip mounted in its package

Fig. 1.6    The inside of a semiconductor clean-room fabrication facility (courtesy GEC Hirst)

1.2 Plot a graph of the overall yield, $Y$, against the number of processing steps, $N$, for a series of values of the yield per stage, $f$, using equation 1.1

1.3 Using the graph of problem 1.2, discuss the problems associated with increasing the complexity of any given process sequence.

1.4 If the number of stages in problem 1.1 is doubled, by how many percentage points does the yield change?

1.5  Calculate the number of 6 mm by 6 mm square integrated circuits that can be made on a 10 cm diameter silicon wafer.

1.6 Calculate the number of 6 mm by 6 mm square integrated circuits that can be made on a 15 cm diameter silicon wafer.

# CHAPTER 2

# Materials for Semiconductor Devices

### Instructional Objectives

*This chapter presents an overview of some of the commonly used semiconductor material growth techniques. You will learn about:*

a. Zone refining techniques.
b. Single crystal growth procedures.
c. Czochralski growth procedures.
d. Float zone crystal growth procedures.
e. Epitaxial crystal growth techniques.
f. Chemical etching techniques.

### Self-evaluation Questions

*Watch for the answers to these questions as you read the chapter. They will help point out the important ideas presented.*

a. Why is zone refining necessary in some base material?
b. How does Czochralski crystal growth differ from float zone crystal growth?
c. Why is float zone silicon usually cleaner than Czochralski?
d. Where is epitaxial growth material used?
e. What layer thicknesses are produced using vapour phase epitaxy?
f. What are the advantages of the advanced technologies of MBE and MOCVD over conventional vapour phase epitaxy?

## 2.1  INTRODUCTION

A key component in the whole of the current activity in semiconductor microtechnology is the production and control of the base semiconducting material, from which devices and integrated circuits are made. Device quality semiconductors are composed of single crystals of high perfection and high purity. The production of such base material demands many years of concentrated research and development to achieve the necessary standard. The problem is clearly illustrated when one considers that William Shockley[†] proposed a field effect transistor structure in 1952, but it was not until 1962 that silicon technology had developed to a point where such devices could be constructed with reproducibility in characteristics from one device to the next. At the present time silicon technology has reached the stage where such reproducibility and reliability is achieved in complex integrated circuits containing up to 1000 000 transistors. This is no mean achievement! Gallium arsenide is another important semiconductor, but here the technology is much more primitive when compared to silicon and we are at the present time witnessing a concentrated effort to raise the standard of its technology so that integrated circuits may be made which possess a very high operational speed resulting from the high electron mobility in this material. Each semiconducting material brings with it its own set of problems which must be solved before a useful device can be fabricated precisely, and with reproducible characteristics. Only then can factory production be contemplated. A further and important consideration that will decide whether a particular material or technique will be used commercially is the cost. In order to establish a new fabrication technique, it is first essential to show that improved performance will result, or alternatively that the new material (or device) has some feature unattainable in another material. Bearing these considerations in mind we will now outline some of the general features of present-day semiconductor technology. The very limited nature of this introductory discussion will, therefore, be selective rather than exhaustive.

## 2.2  CRYSTAL PURIFICATION AND GROWTH

The majority of solids produced in nature are grossly imperfect and impure, a direct result of the haphazard way in which they are formed. Thus the first stages in producing a semiconductor device is the purification of the material and then the growth of high-quality single crystals under controlled and ultra-clean conditions. Techniques of crystal growth have in recent years proliferated, each material having demanded its own variation to overcome its particular problems. A brief description of these techniques will now be presented.

[†] W. Shockley, A unipolar field effect transistor, *Proc. IRE*, **40**, 1365–76 (1952).

### 2.2.1  Zone refining

Prior to the growth of a single crystal of say silicon or germanium or gallium arsenide, the base material must be purified to a degree beyond that attainable by simple methods. For example, in some devices the residual impurity content must not exceed about one impurity atom for every hundred million host atoms! A standard method of purifying materials is zone-refining, illustrated in Fig 2.1(a). The general principle underlying this technique is that the maximum unwanted impurity concentration of the molten state exceeds that of the same material in the solid state. Thus if a rod of polycrytalline material is slowly passed through a heating coil such that a molten zone passes from one end of the rod to the other, many impurities are concentrated in the liquid zone and are swept to the ends of the crystal (which are later discarded). If the refining process

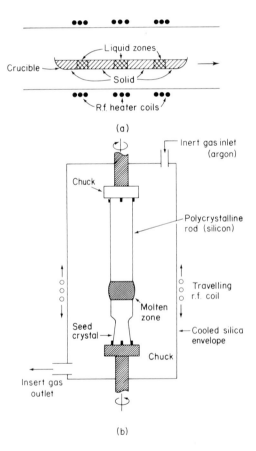

(a)

(b)

Fig. 2.1   (a) Zone refining of material using a horizontal boat; (b) zone refining and crystal growth using the boat-free vertical method

is repeated by multiple passes, then an extremely pure material can be prepared. An adaption of this is the floating zone method. Here the rod is held vertically and a single molten zone is passed vertically from end to end. Surface tension in the liquid zone maintains the liquid between the two solid portions (Fig. 2.1(b)). Since the solid acts as the container for its own melt, contamination from a crucible is avoided.

## 2.2.2 Crystal growth

At the end of the zone-refining process, material of a high purity is obtained. The atoms in this solid, however, are not arranged in the simple periodic structure

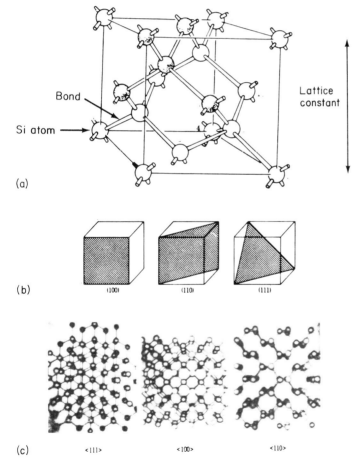

Fig. 2.2 The crystal structure of silicon: (a) the unit cell; (b) the three major crystallographic planes; (c) looking along the three major axes

which is the characteristic of a *single crystal* (Fig. 2.2). In order to change this material into a near perfect crystal, a process called *crystal growth* must be carried out. One way of achieving this is to convert the solid into a liquid and then allow this liquid to be cooled in a controlled way, such that each successive row of atoms is allowed to solidify, by moving into the correct position in the solid. This may be achieved by allowing the new solid to form on the surface of a 'seed crystal'. The seed is a small nucleus of crystal, which itself is highly perfect and hence can nucleate the structure of the new crystal. An alternative process relies on the conversion of the desired atom directly from the gas phase into a solid, and this is termed vapour phase epitaxy. In crystal growth the ideal situation is for the seed to be of the same material as the growing crystal, as is always the case with silicon. However, this is not always possible, particularly so in some of the more recently developed binary (two elements) ternary (three elements) and quaternary (four elements) materials. We will now discuss some of the processes in more detail.

### 2.2.3  Bulk grown crystals: the Czochralski growth process

The growth of a bulk ingot by this process is shown schematically in Fig. 2.3. The material for crystal growth is placed in a crucible and brought to a *melt*,

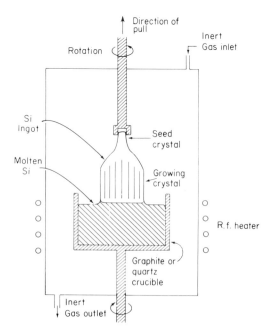

Fig. 2.3   The Czochralski crystal growth process

usually by means of a radio frequency (r.f.) heating coil. The seed is mounted on a chuck which can rotate about a vertical axis and may also be drawn slowly away from the melt as shown in Fig. 2.3.

The procedure at this stage is to place the seed crystal at the surface of the melt and then to draw it away slowly. As the seed is withdrawn it raises a film of liquid, which cools, solidifies and is forced to take on the structure of the seed. The pull is continued smoothly, thus allowing a cylindrical ingot to grow, as is illustrated in the photograph shown in Fig. 2.4(a), which shows an 80 mm diameter gallium arsenide ingot. In the early days of technology only relatively small diameter (30 mm) ingots could be drawn, this constrasting markedly with the large (150 mm) diameter ingots that are readily obtained with current technology. An example of this type of crystal puller is illustrated in Fig. 2.4(b), which shows the 'Melbourn' III-V crystal puller. Note that in this instrument the whole of the growth process is carried out at elevated pressures and the whole system has to withstand these extreme conditions. Another very important practical point to note relates to the diameter of the initial section of the growing crystal. The diameter of the growing crystal is encouraged to decrease initially and then to increase over the first centimetre. This is achieved by starting the pull at a relatively high speed and slowing down the pull rate (i.e. the diameter of the growing crystal is related to the rate of growth). This necking technique allows any dislocation to be terminated in the necked region, resulting in a much higher quality crystal.

Fig. 2.4  (a) A single crystal of gallium arsenide (8 cm diameter) grown by the Czochralski process

Fig. 2.4   (*continued*)   (b) The 'Melbourne' puller used to grow the gallium arsenide crystal (permission of Cambridge Instruments Ltd, England)

The process described above will ideally produce an intrinsic[†] semiconductor. Alternatively, material may be doped p- or n-type by introducing the appropriate p- or n-type doping pellet into the melt. It should also be stressed at this juncture that the growth of a binary solid such as gallium arsenide (GaAs) is much more complex than, for example, silicon and the details of this are the domain of a more advanced book.

---

[†] An intrinsic semiconductor is one which contains no intentionally introduced impurities.

## 2.2.4   Float zone process

This technique is in essence an extension of the zone-refining described earlier and illustrated in Fig. 2.1(b), the difference here being that a seed crystal is clamped up to one end of the polycrystalline rod. The position of the heating coil is such as to produce a liquid zone at one end of the polycrystalline rod supported by surface tension forces. A seed crystal is placed into the liquid zone. The liquid zone is now moved along the rod allowing the rear surface to cool and solidify in the structure of the seed crystal. In order to dope the crystal, the dopant can either be distributed uniformly in the polycrystalline rod, or alternatively it can be inserted periodically into the rod by drilling a sequence of holes along the length and loading these with the dopant material. For the best uniformity, however, certain dopant materials may be included in the surrounding gas. The gas phosphine, for example, can provide the phosphorus needed for n-type doping of silicon. One of the most important advantages of this technique over the Czochralski process is that the molten silicon is not contained in a boat and is thus less susceptible to contamination by unwanted impurities—particularly oxygen. Crystals with very high values of resistivity have been grown using this technique.

## 2.2.5   Epitaxial growth

Bulk-grown crystals of a quality sufficiently good for direct device fabrication are difficult to grow. One way of circumventing this difficulty is to use a polished slice of bulk grown crystal as a foundation and to grow onto this surface a much higher-grade *epitaxial* layer for the actual device structures.

In the process of epitaxial growth, a polished bulk grown sample (i.e. a wafer) is used as the seed for epitaxial growth. The word epitaxial means that the new crystal layer has the same crystal structure as the base wafer. The detailed processes of epitaxial growth are numerous and varied to cater for the needs of different types of semiconductor. A simple crystal such as silicon clearly will differ in its needs from a more complex compound semiconductor such as gallium arsenide. We will therefore confine our attention to principle and not to detail.

## 2.2.6   Liquid Phase Epitaxy (LPE)

The basic principle underlying the process of liquid phase epitaxy is shown in Fig. 2.5(a). A substrate (i.e. seed) crystal is held above a semiconductor melt and then dipped into it. As the substrate is withdrawn from the melt to the cooler region of the furnace, the molten film covering the surface will form an epitaxial crystalline layer provided the rate of cooling (i.e. the withdrawal rate) is carefully controlled.

Practical systems, although more complex, rely on the same basic principle. For example, Fig. 2.5(b) shows the apparatus used for liquid phase epitaxy of the semiconductor material gallium arsenide. In this sophisticated system, the semiconductor growth is held in a graphite boat located in a quartz tube with very pure hydrogen gas flowing through the system. Also shown in this figure is a cross section of the graphite boat and its melts with the gallium arsenide source crystal at the bottom and top of the melts. The bottom section is a slider

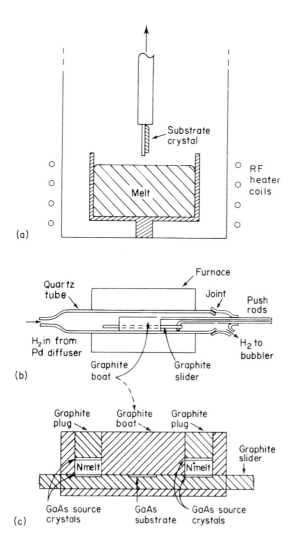

Fig. 2.5  (a) A schematic of the liquid phase epitaxial (LPE) growth system; (b) an LPE system for gallium arsenide; (c) an enlargement of the boat assembly (permission of Cornell University)

that is used to slide the bottom source crystal or the growth substrate into position under the melt. The pure gallium melt dissolves increasing amounts of arsenic from the gallium arsenide source crystal as its temperature is raised. Once in the gallium melt, the arsenic can be removed by cooling the melt in contact with the substrate single crystal (i.e. by sliding the crystal to be used for growth under the melt). This forces gallium arsenide to grow on the substrate surface. Variations of this technique may be used for other crystal systems.

### 2.2.7  Vapour Phase Epitaxy (VPE)

As in the case of liquid phase epitaxy, there are many variations of the basic technique. Figure 2.6 illustrates the principles of the silicon growth scheme. The substrate of single crystal is placed inside a quartz furnace tube and held at approximately 1250 °C. Silicon tetrachloride vapour carried in a stream of hydrogen gas is passed through the furnace. Inside the heated region the chemical reaction

$$SiCl_4 + 2H_2 \rightarrow Si + 4HCl \tag{2.1}$$

takes place. The silicon produced in this way is deposited and forms a single crystal on the substrate surface. Very pure epitaxial films can be fabricated by controlling the purity of the chemicals; alternatively, the crystal can be deliberately doped n- and p-type by first bubbling the hydrogen through a weak solution of phosphorus trichloride (n-type) or boron trichloride (p-type). This procedure produces epitaxial layers 2–20 $\mu$m thick with a resistivity that is controllable to within 5% from one crystal to the next.

Another example of a VPE system, used for growing compound semiconductors such as GaAs and InP, is shown in Fig. 2.7. Consider the growth of GaAs. Gallium is used as the source material and held at 800 °C. A crust must form over the entire surface before crystal growth begins to ensure arsenic saturation. Before crystal growth begins, a back etch (i.e. dissolving) of the surface is carried

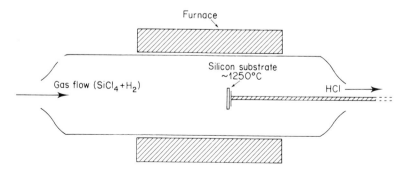

Fig. 2.6   A vapour phase epitaxial (VPE) system for silicon

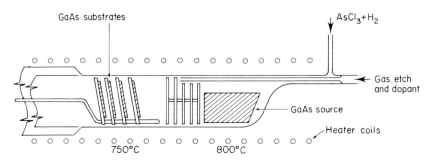

Fig. 2.7   A VPE system for gallium arsenide (reproduced by permission of Plessey Company plc.)

out by running the system at the relatively high temperature of 850–900 °C. Thus assured of a good growth surface, the epitaxial layer is deposited at 710–750 °C. Highly doped material may be grown by moving the material upstream and lightly doped by moving it downstream in the $H_2S$ gas. By balancing the arsenic trichloride flow relative to the $H_2$ flow, it is possible to control the ambient net donor density over a significant range, including the growth of very pure undoped (or intrinsic) materials.

To conclude our discussion we will discuss two relative newcomers to the crystal growth business, *M*olecular *B*eam *E*pitaxy (MBE) and *M*etal *O*rganic *C*hemical *V*apour *D*eposition (MOCVD).

### 2.2.8   Molecular Beam Epitaxy (MBE)

The basis of this technique is to allow a beam of the desired constituent atoms to fall upon, and stick to, a desired substrate held at an elevated temperature in ultra-high vacuum chosen such that it allows the best crystal to grow on the substrate. In the case of compound semiconductors, separate cells for each of the component elements provide the atoms required for growth. The source of the molecular beams, called *Knudsen* effusion cells, are in essence heated enclosures containing the elements required for the molecular beam. The elevated temperature ensures the desired high vapour pressure and a suitable orifice allows the beam to emerge in the desired direction. In practice the difficulty in growing a multi-atom compound semiconductor, such as gallium-indium-arsenide (GaInAs), is to ensure that each of the sources is at the correct temperature to allow the molecular beams arising from the vapour pressure in the cell to form a stoichiometric solid. A schematic diagram of a typical molecular beam epitaxial system is shown in Fig. 2.8. Basically, this is a very expensive technique, since all the crystal growth takes place in an ultra-high vacuum chamber. This also has its advantages, however, since a range of auxiliary facilities may be placed inside the chamber. An electron gun allows electron diffraction patterns of the crystal

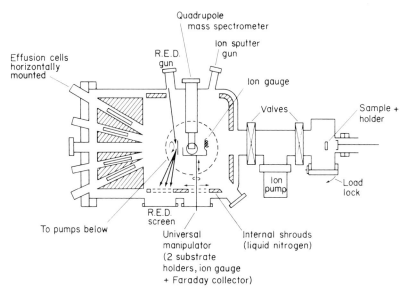

Fig. 2.8  A schematic layout of an MBE system (permission of Cornell University)

to be monitored throughout the growth and the concentration of the molecular beam constituents may be monitored precisely using a quadrupole mass spectrometer. The doping of epitaxial layers presents no real difficulties provided that they have high sticking coefficients to the epitaxial layer concerned (i.e. they have a high probability of sticking to the growth surface).

### 2.2.9  Metal Organic Chemical Vapour Deposition (MOCVD)

A schematic diagram showing the basic prinicples of this technique is shown in Fig. 2.9. In order to illustrate these principles, consider the growth of gallium arsenide. The group III element Ga is transported with the $H_2$ gas to the reactor. Similarly, the group V element As is transferred and meets the gallium component just before the entrance to a cold walled reactor. Inside the reactor the two gases pass over an inductively heated substrate, where the reaction

$$Ga(CH_3)_3 \uparrow \; + AsH_3 \uparrow \; \xrightarrow[\text{substrate}]{\text{hot}} \; GaAs(Solid) + 3CH_4 \uparrow \qquad (2.2)$$

occurs only on the heated substrate. A temperature range of 600–800 °C is typical of those required for the reaction. If doped semiconductors are required,

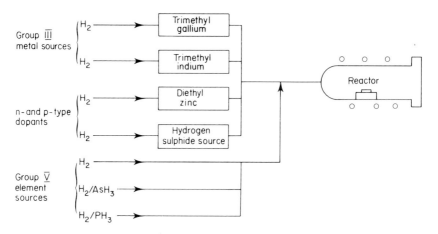

Fig. 2.9   A schematic layout of the gas input lines for metal organic CVD (MOCVD) of gallium arsenide (permission of British Telecom)

then p- or n-type dopants can be introduced into the gas in the manner shown in Fig. 2.9. This technique differs from vapour phase epitaxy in the important respect that the growth reactor vessel is cold, only the crystal seed wafer being selectively heated to initiate the crystal growth.

The latter two techniques have much in common. They are both best suited to the growth of special profiles which are not readily achieved by the other established technologies. Such profiles are used, for example, in special microwave and optoelectronic device structures. The two techniques are essentially of a non-equilibrium nature and can be used to achieve very rapid changes in doping profile. Both methods are being studied in many laboratories at the present time and the current achievements show very encouraging trends for future applications.

## 2.2.10   Chemical vapour deposition

The device fabrication process frequently requires the deposition of thin insulating layers onto the semiconductor surface. This can be achieved by sputtering as described earlier, or thermal oxidation as in the special case of silicon (Chapter 4). Another important class of deposition is chemical vapour deposition (CVD) which is regularly used for silicon dioxide, silicon nitride, and polycrystalline silicon (polysilicon). The basic principle of the technique is to react a material at high temperature and to deposit it on the crystal surface with the aid of a carrier gas. The vapour phase epitaxial technique is a good example of this process.

For silicon dioxide, deposition can be achieved via the reaction of silane at 200–500 °C using an $N_2$ carrier gas, i.e.

$$\text{Silane } (Si(H_4)) + 2O_2 \rightarrow SiO_2 + 2H_2O. \tag{2.3}$$

For silicon nitride at 750–850 °C the reaction is

$$3SiH_4 + 4NH_3 \rightarrow Si_3N_4 + 12H_2, \tag{2.4}$$

and for polysilicon a reaction takes place at 500–800 °C:

$$SiH_4 \rightarrow Si + 2H_2. \tag{2.5}$$

These films can be of an amorphous nature, where the arrays of silicon atoms have no short-range order and are entirely random, or polycrystalline where columnar grains are formed which are themselves single crystals of silicon, but which have random orientation to adjacent grains. The main factors governing which type is formed is growth temperature and pressure. The transition temperatures between amorphous and polycrystalline films are around 550 °C when boron is the impurity being incorporated, and near 630 °C when the relevant impurity is phosphorus. While deposition often takes place at atmospheric pressure, improved film quality can be obtained by allowing the reaction to take place at low pressures in what is known as a low-pressure CVD (LPCVD) reactor.

Polysilicon is frequently used in MOS integrated circuits where it has been found to be valuable as an additional conducting layer when heavily doped with boron (p-type) or phosphorus (n-type) to render it of low resistance. A further discussion on deposited films is introduced in Section 4.9.

### 2.2.11  Growth of heterojunctions

Silicon devices and integrated circuits are the principal subject of this book. However, there have been considerable efforts to study and employ other semiconductors. A primary reason for this activity arises from the desire to vary the material band gap, a feature of particular importance in, for example, optical devices. In recent years another important device design technology has proved of considerable importance, that is band gap engineering for the fabrication of heterojunction devices. It is really the research on advanced growth techniques such as Molecular Beam Epitaxy (MBE) and Metal Organic CVD (MOCVD) that is responsible for the rapid advances in this technology. These crystal growth techniques (Sections 2.2.8 and 2.2.9) enable epitaxial layers to be grown with variations in material parameters. Doping impurity atoms and even compositional changes can be varied rapidly during growth. Thus a heterojunction between two dissimilar materials such as GaAs and AlGaAs can be grown. Figure 2.10(a), (b) and (c) show three examples which illustrate the strength of this new technology. In the first

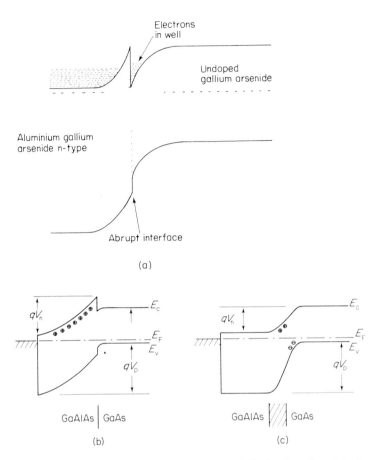

Fig. 2.10   Illustrating the band diagrams at heterojunction interfaces: (a) abrupt junction between undoped GaAs and n-type AlGaAs; (b) abrupt junction between p-type GaAs and n-type AlGaAs; (c) a graded heterojunction for the same stucture as shown in (b)

example we have an abrupt junction between an undoped GaAs and n-type AlGaAs. The abrupt junction results in the step discontinuity in the band diagram and the formation of a small potential well in the conduction band of the GaAs. To make the system more complex a thin region ($\simeq 200$ Å) of AlGaAs adjacent to the interface is undoped. Such a growth potential allows the interface energy band to be engineered in order to produce novel semiconductor devices (Chapter 9).

The second and third examples are also heterojunctions between GaAs and AlGaAs. In Fig. 2.10(b) the junction is abrupt and is formed between an n-type AlGaAs and a p-type GaAs.

Finally the third example shown in Fig. 2.10(c) is a heterojunction where the

AlGaAs is graded. That is the Al concentration is increased slowly from the value of zero in GaAs to a value of about 30% in the AlGaAs and then remains constant. In this example the doping is also changed from n-type in the AlGaAs to p-type in the GaAs, a very sophisticated junction structure. The system described above (AlGaAs/GaAs) is just one of many possibilities currently being grown.

One of the major restrictions in the growth of a heterojunction is the need to ensue that the two materials being grown have nearly the same separation between their atoms. This is referred to as 'lattice matching'. If these interfaces are not lattice matched strain is built up in the growing layer which can only relieve itself by the formation of large numbers of dislocations and hence the production of defective materials. Figure 2.11 is important to the crystal grower since it tells him which materials have the same lattice constant (i.e. atomic separation). From this figure we may see that GaAs and AlGaAs have the same lattice constant and hence are lattice matched and suitable for heterojunction growth.

In situations where the two materials are not lattice matched it is still possible to grow good quality layers, provided the grown layer is fairly thin (less than 100 Å). In these circumstances the strain arising from the mismatch results in a deformation of the grown (thin) layer and is called a strained lattice. If the layers become too thick the strain will release itself by forming dislocations. Such thin layers have proved to be of value to the device engineer.

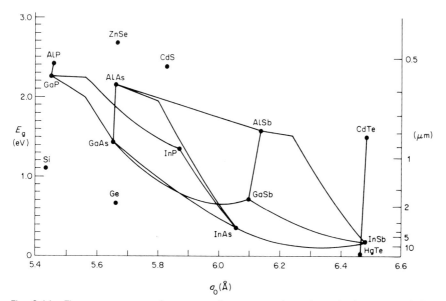

Fig. 2.11   The energy gap of a range of compound semiconductors versus lattice spacing. Note that GaAs→AlGaAs→AlAs are along a near vertical line, thus indicating little change in lattice constant for this important range of compounds

### 2.2.12    Plasma deposition

One of the main difficulties with the use of CVD techniques is the relatively high temperatures involved. At the lower deposition temperatures the layer quality, particularly the insulators, is not good enough for device fabrication. Further-more, the relatively high temperatures cannot be tolerated when metallizations (Chapter 6) such as aluminium are present on the surface. In plasma deposition the active species is present in a gas plasma enabling reactions to take place at around room temperature which would not occur outside the plasma. This technique is being used to produce good quality silicon nitride, which is very valuable in silicon and gallium arsenide technology. The semiconductor wafer is placed inside the plasma vessel and the (gaseous) active species is allowed to react and deposit on the surface.

## 2.3    WAFER PREPARATION

In order to fabricate devices or integrated circuits, the starting material, the bulk crystal, must first be sliced into wafers. Devices may then either be fabricated directly in the bulk crystal, or indirectly by using the bulk crystal as a seed for epitaxial growth as was discussed earlier. In many applications the bulk crystal rod is ground into a cylinder shape to ensure a consistency in the diameters of the eventual wafers—this is important for production-line handling in subse-quent IC processing. Automation does depend critically on handling wafers with identical dimensions. To ensure this situation the boule is ground on a lathe to form a constant diameter cylinder. The orientation of the surfaces normal to the

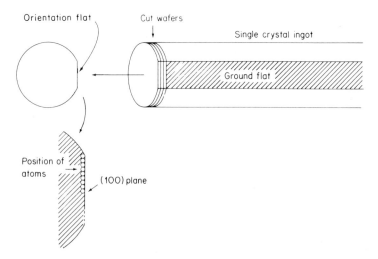

Fig. 2.12    Illustrating a single ingot of silicon with the ground flat aligned to a (100) plane

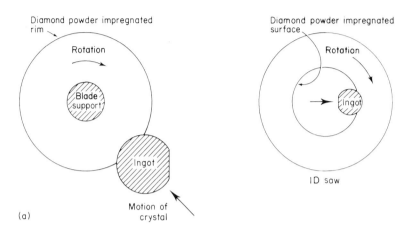

(a)

(b)

Fig. 2.13   (a) Illustrating the microslice annular saw process; (b) photograph of a real saw (permission of Cambridge Instruments Ltd)

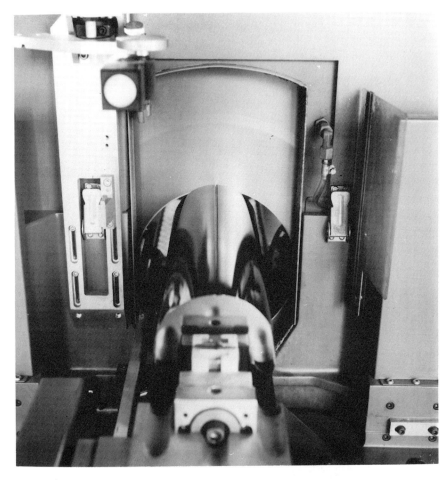

Fig. 2.13  (*continued*)  (c) A commercial annular saw for cutting silicon ingots. The silicon ingot is shown being stepped into the face of the blade for cutting (Courtesy Wacker-Chemitronic)

cylinder axis is fixed by the seed crystal (and for silicon, gallium arsenide, and indium phosphide it is usually (100) and occasionally (111)). The orientations within the plane must be fixed by X-ray diffraction. Once the geometry of the cylinder is determined, a flat is ground along its length to provide a permanent reference on each of the final wafers (Fig. 2.12). This step is very important since the final crystal wafer will have preferential cleavage planes which may be used for breaking the whole wafer into its component devices or ICs. The edges of the separate ICs must be aligned so that they are parallel to the cleavage planes. The cylindrical rod is then sliced into thin parallel wafers as is illustrated in Fig. 2.12.

For this delicate cutting process, a special diamond-impregnated wheel is used (Fig. 2.13). A fine ring shaped blade is coated with diamond powder (i.e. very fine particles of diamond) which are sufficiently hard to saw away the crystal rod.

Two forms of diamond saw may be used, one with the cutting edge as the outer surface or, alternatively, one with the cutting edge on the inside. These are illustrated in Fig. 2.13(a). The mechanical rigidity of the saw with the internal cutting edge permits the use of a thinner blade wheel which does not ripple and hence produces a thinner cut with less wasted semiconductor between slices.

At this point in the procedure, the sawing process leaves wafers with some marks which must be removed by using a suitable etchant (i.e. a chemical solution which will slowly dissolve the surface region of the semiconductor). This chemical polishing process will remove the damage from the immediate cutting area. Unfortunately, however, the cutting damage will propagate some distance into the slice and must be removed if a surface is to be presented for crystal growth or for direct device fabrication. This cannot however be done by continued etching, primarily because etching will increase rather than decrease surface topographical irregularities. To avoid this difficulty a process of mechanical/chemical polishing is used (Fig. 2.14). The wafer is mounted on a rigid arm

Fig. 2.14   A commercial wafer polishing facility (courtesy Wacker-Chemitronic)

suspended over a rotary polishing disc. Polishing takes place by impregnating the disc with a chemical etchant which slowly dissolves the semiconductor surface. This technique ensures that the wafer surface is polished into an optically flat finish. As a final step, the wafer is subjected to a short conventional etch to remove any residual damage and cleaned to remove any residual contamination ready for device processing or alternatively as the substrate (seed crystal) in an epitaxial growth process.

## 2.4  CHEMICAL ETCHING

In the preceding section chemical etching or polishing was introduced as a means of removing surface damage resulting from the sawing of the wafers. Surface etching has however a much wider use in microtechnology and this will now be discussed. One important use of etching is in the successive diffusion (or ion implantation) stages in the making of an IC. Here it is necessary to cut diffusion windows in the protective surface oxides, to remove oxides completely or to cut windows for metallization purposes. In the case of silicon devices the oxide used is thermal silicon dioxide ($SiO_2$). A suitable etchant for this is hydrofluoric acid (HF) or a mixture of this acid with $NH_4F$. In addition to this, a number of other etches have been developed for specific applications. Some of the more common solutions are summarized in Appendices 2 and 3. Etches are also needed to remove selectively, metal overlayers, the specific etches depending very much on the material that requires removal.

Returning to the crystal surface itself, a number of specialist etches have been developed over the years. Some etches produce Mesa structures, other etches are used for selectively decorating dislocation (in order to assess the perfection of the starting material) and some etches are used for decorating pn junctions. In this last category are chemicals that act differently on p- and n-type material. By selectively etching on one side of a junction, its actual position can be made visible and measured (Appendix 1). In the second category the etch concerned acts more rapidly on the less ordered material in the vicinity of a dislocation and hence produces a small pit at the point where the dislocation reaches the surface.

### 2.4.1  Anisotropic etches

Recently a number of etches have evolved which are capable of selectively producing a well-defined etch trough with either a V- or a U-shaped cross section. Such etches are called *anisotropic etches* and their successful use in device fabrication reflects the advanced state of development of silicon technology.[†] To illustrate the principles of the technique consider one 'V'

[†] K. E. Bean, Anisotropic etching of silicon, *IEEE Transactions on Electron Devices*, ED-25 No. 10, pp. 1185–93, October 1981.

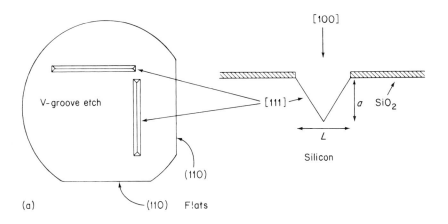

Fig. 2.15 (a) Illustrating the 'V' groove anistoropic etching process; (b) V-grooves in silicon (permission of Siliconics Ltd, Swansea)

groove etching technique. The basis of operation of these etches is related to the variation in packing density of the different planes in the silicon (or gallium arsenide) crystal structure. Figure 2.2 illustrates this point and defines the three major planes (100), (110), and (111) in the crystal. From this figure it is readily seen that the atomic packing density and available bonds decrease as we go from (111) to (100) to (110) and hence selective etches are able to remove atoms in some directions more readily than others.

This principle is illustrated in Fig. 2.15 which shows an oxide coated (100) silicon slice with two long window cuts parallel to the (110) directions. When such a sample is placed in an orientation dependent etch (such as KOH, normal propanol, and $H_2O$), the etching proceeds to the $\langle 100 \rangle$ direction until the etch front hits the two $\{100\}$ planes intersecting the (100) plane at the edge of the

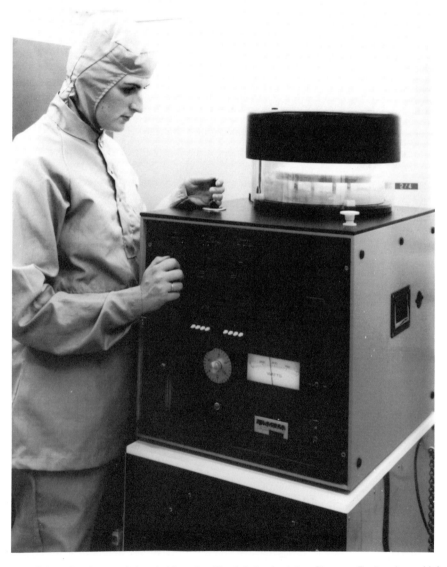

Fig. 2.16   A plasma lab etching facility fabricated by Plasma Technology Ltd (Courtesy GEC Hirst)

mask opening and then etching stops with the resulting 'V' groove. The ratio $a/L$ of the etch depth $a$ to oxide opening $L$ ($a/L \sim 0.707$) is fixed by the photolithographic opening $L$ and the geometry of the crystal structure.

As an example of the application of such an anistropic etch Fig. 8.4(b) shows the structure of a V groove channel FET in silicon.

### 2.4.2  Plasma etching

The wet chemical etching processes described above have limitations when very fine patterns have to be produced. This is in essence due to the viscosity limit of the liquids needed to bring in new active radicals and the removal of the reaction products of the etch. These limitations can be overcome by plasma etching. In this case the etching is achieved by using the active species found in the plasma to attack the semiconductor surface.

The earliest plasma process used oxygen to remove photoresist. As an example of a plasma suitable for etching silicon, silicon oxide or silicon nitride the complex reactions resulting from $CF_4$ gas can be used. The details of practical systems are complex and the interested reader may obtain further details in Ref. 5 in Appendix 6. Figure 2.16 shows a plasma etching facility with the etching chamber.

### 2.4.3  Ion beam etching (ion milling)

Ion milling is in essence the process whereby the surface of a solid is slowly eroded by using an ion beam. A collimated beam of ions with energies in the range 500 eV to 1 keV (Fig. 2.17) are directed onto the surface of the solid. These ions can strike the near-surface atoms of the solid and in the process communi-

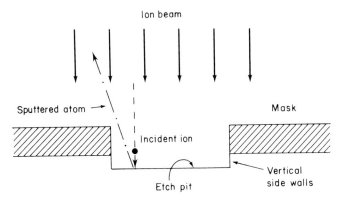

Fig. 2.17  Illustrating the ion milling process

cate to them sufficient energy to enable them to leave the surface. This process is called *sputtering*. The use of argon ions as the sputtering beam is a convenient working gas. One important advantage of this technique is that it produces nearly vertical etch pits with no problems of undercutting of the mask. This is important for future developments in submicron structures.

## Problems

2.1    What is meant by the term 'single crystal'?

2.2    How does MOCVD differ from VPE?

2.3    What is the atomic spacing between two silicon atoms?

2.4    What is the atomic spacing between two germanium atoms?

2.5    What is the atomic spacing between two gallium arsenide atoms?

2.6    What is meant by the term 'polysilicon'?

2.7    Why must a semiconductor wafer be etched after wafer cutting?

2.8    What type of etches are used to produce V- or U-shaped troughs in silicon?

2.9    What is meant by the term 'lattice matching' in heterojunction growth?

2.10   What are the advantages of plasma etching over conventional wet etching?

# CHAPTER 3

# Impurity Doping in Semiconductors

## Instructional Objectives

*This chapter introduces the concept of intrinsic and extrinsic semiconductors and discusses the important ways in which impurities are introduced into semiconductors.*

a.  Explain the terms intrinsic and extrinsic semiconductors.
b.  Describe the role of majority and minority carriers in semiconductors.
c.  Describe doping techniques and calculate impurity profiles arising from diffusion and ion-implantation.

## Self-evaluation Questions

*Watch for the answers to these questions as you read the chapter. They will help point out the important ideas presented.*

a.  Why is the intrinsic concentration of carriers for most semiconductors not zero at 300 K?
b.  Why does silicon become p-type when boron is introduced as an impurity?
c.  Why are impurities such as copper troublesome in some semiconductors?
d.  What are the two most common methods used to introduce impurities into semiconductors?
e.  Why does ion implantation doping offer the potential of direct-write patterning?

## 3.1  BASIC CONCEPTS

### 3.1.1  Intrinsic semiconductors

Semiconducting material may, for convenience, be divided into two categories, intrinsic and extrinsic[†]. Consider first the case of intrinsic semiconductors. In their pure form these materials are insulators at very low temperature but begin to conduct electricity as they are heated, by virtue of their relatively narrow band gaps. Figure 3.1 shows an energy band diagram of such an intrinsic semiconductor. There are no fundamental differences between intrinsic semiconductors and insulators, the division is somewhat artificial since by convention intrinsic semiconductors are defined as insulators with band gaps of less than about 2 eV. The reason for this division may be understood if one considers the statistical process taking place within solids when thermal energy (heat) is supplied to the crystalline atoms. When this energy is communicated to the valence electrons, some will (statistically) gain sufficient energy to cross the forbidden gap and produce both electrons (n) and holes (p) which may then move under the influence of an applied electric field to produce current. A full mathematical treatment[‡] of the problem establishes that the number of electron-hole pairs at a temperature T is given by the equation

$$p = n = n_i = G \exp\left\{-\frac{E_g}{2kT}\right\}, \tag{3.1}$$

where $G$ is a constant which varies slowly from material to material. In the table below we show the values $n_i$ at 300 K for some important semiconductors.

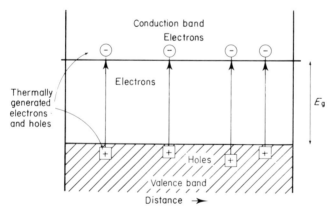

Fig. 3.1   A simple band diagram of a semiconductor

[†]For further disussion see Ref. 6 in Appendix 6.
[‡]See Ref. 6 (pp. 17–27), Appendix 6.

| Semiconductor | Ge | Si | GaAs |
|---|---|---|---|
| $E_g$ (eV) | 0.66 | 1.12 | 1.42 |
| $n_i$ (300 K)(m$^{-3}$) | $2.4 \times 10^{19}$ | $1.45 \times 10^{16}$ | $1.79 \times 10^{12}$ |

From this table we see that at room temperature $n_i$ varies from approximately $10^{19}$ m$^{-3}$ for germanium to $10^{12}$ m$^{-3}$ in gallium arsenide. Thus the number of electrons and holes that are available to transport electricity decreases by seven orders to magnitude for the relatively small change in band gap from 0.66 eV to 1.42 eV. As a consequence of the very strong exponential dependence of $n_i$ on band gap $E_g$, a semiconductor with a band gap of greater than about 2 eV is deemed an insulator at room temperature, because of the low value of $n_i$ and hence its low conductivity.

Note that the conductivity $\sigma$ is given by

$$\sigma = nq\mu_n + pq\mu_p$$
$$\sigma = n_i q(\mu_n + \mu_n) \tag{3.2}$$
$$\sigma = q(\mu_n + \mu_p)G \exp(-E_g/2kT)$$

hence

$$\ln \sigma = \ln (\text{constant}) - \frac{E_g}{2k}\left(\frac{1}{T}\right). \tag{3.3}$$

Here we have assumed that the temperature dependences of $G$ and the mobilities $(\mu_n + \mu_p)$ are negligible.

Thus a plot of log $\sigma$ versus $1/T$ is a straight line (Fig. 3.2) with a slope of $-E_g/2k$. This is a valuable experimental method of determining the band gap $E_g$.

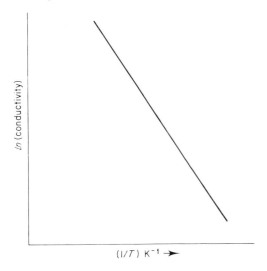

Fig. 3.2 A plot of ln (conductivity) versus $(1/T)$ for an intrinsic semiconductor

### 3.1.2    Extrinsic semiconductors

An *extrinsic* semiconductor is so named because its conductivity is produced by the *external* influence of specific impurity atoms. These are atoms with a valence of either three or five, substituted in place of regular silicon (or germanium) atoms. It is important to stress that these impurities must sit on a regular silicon site if they are to be electrically active (Fig. 3.3(a,b)) and not an interstitial site as shown in Fig. 3.3(c).

Impurity atoms such as phosphorus or arsenic, which have a valence of five and are therefore termed donor atoms, have one more electron than is required for the covalent bonding. This excess electron can readily be torn away from its parent atom, allowing it to respond freely to an electric field. This is reflected by the relatively small electron binding energy $\Delta E$ which is required to free the electron, that is to raise it from the defect level to the conduction band. $\Delta E$ is very much less than $E_g$, typically of the order of 0.05 eV for practical donor atoms. The band picture corresponding to this situation is shown in Fig. 3.4(a). This type of material is called n-type because the conductivity is carried by the negative charge carriers—electrons.

The alternative situation resulting from the introduction of an impurity with a valence of 3 such as boron or aluminium in silicon is shown in Fig. 3.3(b). In this case the covalent bonding scheme is deficient of an electron and if such an atom can attract (i.e. accept) an electron from one of its neighbours it will produce a 'hole' and hence result in conduction. Thus these are called acceptor impurities. In this case the conductivity will be by the positive holes and hence the material is termed p-type.

If an n-type semiconductor is doped with $N_D$ atoms m$^{-3}$ where $N_D \gg n_i$, then there will be $N_D$ electrons per m$^3$ in the conduction band (i.e. $n = N_D$) and the conductivity will be

$$\sigma_n = nq\mu_n + pq\mu_p$$
$$\simeq N_D q\mu_n + 0$$

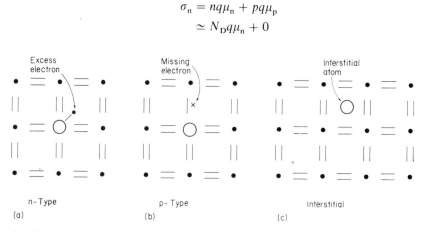

(a) n-Type    (b) p-Type    (c) Interstitial

Fig. 3.3    A schematic picture illustrating (a) donor, (b) acceptor, and (c) interstitial atoms in silicon

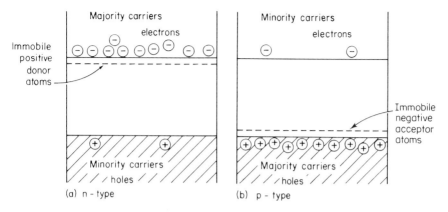

Fig. 3.4   Simple band digrams of (a) n-type and (b) p-type silicon

The number of holes $p$ can be ignored for the following reason. As the number of electrons increases above $n_i$, there is an increase in the probability that an electron and hole will collide and annihilate each other. The technical word for this is *recombination*. As a consequence of this process when a doping $N_D$ is introduced, $n$ increases and $p$ decreases. A formal mathematical treatment of this phenomenon results in the *law of mass action*, i.e.

$$pn = n_i^2. \tag{3.4}$$

Since

$$n = N_D$$
$$p = n_i^2/N_D. \tag{3.5}$$

Consider the case of n-type silicon at 300 K with $n_i \simeq 1.45 \times 10^{16}$ m$^{-3}$.

A typical value of doping for semiconductor device application would be $10^{23}$ m$^{-3}$, and hence

$$p \simeq (10^{16})^2/10^{23} \text{ m}^{-3}$$
$$p \simeq 10^9 \text{ m}^{-3}$$

justifying the assumption that $p \ll n$ and can be ignored.

The same arguments may be used for p-type material and here we write

$$\sigma_p \simeq pq\mu_p$$
$$= N_A q\mu_p$$

where

$$n = n_i^2/N_A \ll p. \tag{3.6}$$

As a result of the small values of $\Delta E$ for these donor and acceptor atoms it requires only a quantity of thermal energy to take away the donor electrons (n-type) or attract a valence electron (p-type). Consequently, these centres become electrically active at low temperatures, typically 20–40 K. Conversely, of course these dopants become electrically inactive at $T = 0$ K. Thus the conductivity versus temperature curve takes the form shown in Fig. 3.5 where $T_F$ is the freeze out temperature—so called because the carriers are frozen onto their parent donor (or acceptor) atoms below this temperature. The conductivity rises abrubtly at $T_F$ and to a first approximation remains constant at a value of $\sigma = N_D q \mu_n$ for n-type ($\sigma = N_A q \mu_p$ for p-type). This is only an approximation since the mobility itself is a function of temperature and decreases slowly as the temperature is increased, as shown by the broken line in Fig. 3.5.

A very important point which must not be forgotten is the constant presence of intrinsic carriers ($n_i$) resulting from the finite temperature of the semiconductor. $n_i$ is strongly temperature dependent and at some elevated temperature $T_i$, $n_i$ will exceed $N_A$ (or $N_D$). At this temperature the semiconductor will revert to being an intrinsic one (Fig. 3.5) and as we will see later this represents an upper temperature limit for the operation of semiconductor devices. Note also that for a given semiconductor $T_i$ will be dependent on the doping density $N_D$ (or $N_A$).

Up to this point the discussion has concerned itself with only one type of dopant: p-type or n-type. It is interesting to speculate on what may happen if the material contains a mixture $N_A$ of p-type and $N_D$ of n-type dopants. For argument's sake let us assume that $N_D$ is greater than $N_A$. The donors will

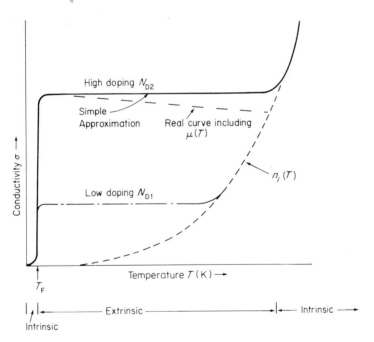

Fig. 3.5   A schematic plot of conductivity versus temperature (K) for an extrinsic semiconductor

donate all their electrons to the conduction band and the acceptor atoms holes to the valence band. In these circumstances the increase in both electron and holes will result in enhanced recombination, thus removing both electrons and holes in the process. A simple view of the situation would be to say that some of the electrons from the donors have passed indirectly onto the acceptor atoms levels. As a result of this process the net number of electrons is $n = N_D - N_A$ and *not* the sum of the two components $N_T = N_D + N_A$. Generalizing this process the following possibilities may be identified:

(a) $N_D > N_A$ and $(N_D - N_A)$ gives a net electron concentration—n-type
(b) $N_A > N_D$ and $(N_A - N_D)$ gives a net hole concentration—p-type
(c) $N_A = N_D$ and $(N_D - N_A)$ is zero and the material appears to be intrinsic!

Case (c) is very important as all the external carriers neutralize each other and the only carriers remaining in the semiconductor are the intrinsic electrons and holes (i.e. $n_i = n = p$). In these circumstances the material is said to be *compensated*. This turns out to be a very useful practical way of producing a 'quasi-intrinsic' semiconductor. For example, if after an extensive purification of a solid (Chapter 2) the net residual defects are p-type with a given density $N'_A$ m$^{-3}$, then crystal is regrown with exactly $N'_D$ ($= N'_A$) donors added intentionally to the lattice. The semiconductor is compensated and appears to be instrinsic. In a material such as GaAs with $n_i \simeq 10^{12}$ m$^{-3}$, true instrinsic material is not available and the current practice is to grow the best material that technology can achieve (which is a net donor density of around $5 \times 10^{20}$ m$^{-3}$, eight orders of magnitude too large!!!) and to compensate with this number of acceptors.

At this juncture it is important to highlight another problem related to doping. If a semiconductor contains $N_D$ donors and $N_A$ acceptors, where for argument we let $N_A > N_D$, the net carrier density is

$$p = N_A - N_D \qquad (3.7)$$

and the conductivity

$$\sigma = (N_A - N_D)q\mu_p. \qquad (3.8)$$

Consider now the microscopic origin of the mobility $\mu$. This is defined as the mean drift velocity of the electrons (or holes) per unit electric field.

If the mean drift velocity of the electrons due to the electric field $F$ is $u$, then

$$\mu = \frac{u}{F}.$$

The electrons (or holes) are at all times undergoing a random walk process (Fig. 3.6) undergoing frequent collisions with atoms at a mean distance (mean free path) $\lambda$ and with a mean time between collisions $\tau$. A rigorous

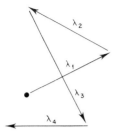

Fig. 3.6  A schematic diagram showing the random motion of a charge carrier (electron or hole) in a semiconductor

treatment of the problem[†] shows that the mobility $\mu$ is given by

$$\mu = \frac{q\tau}{m^*},\tag{3.9}$$

where $m^*$ is the effective mass of the charge carrier. Now it must be borne in mind that any donor or acceptor atom will be charged either negative (acceptor atoms) or positive (donor atoms) and they provide effective centres for scattering electrons and holes. Consequently, large numbers of donors and acceptors will both decrease $\lambda$ and $\tau$, and hence decrease the mobility $\mu$. In these circumstances it is the total number of scattering centres $N_A + N_D = N_T$ that controls the mobility $\mu$ and not the difference, as in the conductivity. An important consequence of this, for example, is that a compensated semiconductor appears to be intrinsic if its conductivity is measured, but can have a mobility much lower than a true intrinsic semiconductor.

This is illustrated for silicon in Figs. 3.7 and 3.8. Figure 3.7 shows the relationship between $\mu_n$ and $\mu_p$ and doping density $N_A$ or $N_D$. Figure 3.8 shows how $\sigma_n = N_D q\mu_n(N_D)$ and $\sigma_p = N_A q\mu_p(N_A)$ vary $N_D$ with $N_A$ in circumstances where there is only one type of carrier $N_D$ or $N_A$.

In the more general case the following procedure must be adopted:

(a)    $N_T$ can be used to determine $\mu$ through Fig. 3.7
(b)    $N_D - N_A$ yields the net carrier density $n_c$ and hence

$$\sigma = n_c q\mu(N_T).$$

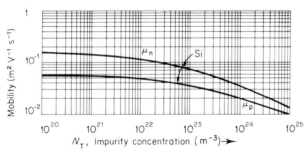

Fig. 3.7    A plot of mobility $\mu_n$ and $\mu_p$ versus total impurity concentration $N_T$ for silicon (after Sze (Ref. 6, Appendix 6))

[†]  See Ref. 6 in Appendix 6 for further details.

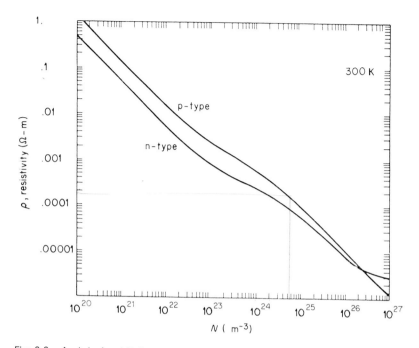

**Fig. 3.8** A plot of resistivity $\rho$ versus doping density $N(N = N_A$ or $N_D)$ for n- and p-type silicon

## Example

Determine the mobility and conductivities of the two silicon crystals

(a)   with $N_D = 5 \times 10^{21}$ and $N_A = 1 \times 10^{21}$ m$^{-3}$
(b)   with $N_D = 2.5 \times 10^{22}$ and $N_A = 2.1 \times 10^{22}$ m$^{-3}$.

## Solution

(a)   $n = 5 \times 10^{21} - 10^{21}$
$\quad = 4 \times 10^{21}$ m$^{-3}$
$\quad N_T = 6 \times 10^{21}$ m$^{-3}$

and hence

$\mu = 0.11$ m$^2$ V$^{-1}$ s$^{-1}$
$\sigma = q(4 \times 10^{21}) \times 0.11$
$\sigma = 70.4$ S m$^{-1}$

(b)   $n = 2.5 \times 10^{22} - 2.1 \times 10^{22}$
$\quad = 4 \times 10^{21}$ m$^{-3}$
$\quad N_T = 4.6 \times 10^{22}$ m$^{-3}$

and hence

$\mu = 0.075$ m$^2$ V$^{-1}$ s$^{-1}$
$\sigma = q(4 \times 10^{21}) \times 0.075$
$\sigma = 48$ S m$^{-1}$

Thus although the net electron carrier density is the same in both materials the mobilities differ and hence the conductivities differ.

### 3.1.3 Unwanted impurities

The impurities described above are the simple and technologically important ones. The behaviour of other impurity atoms in silicon, such as silver, copper, or gold, is much more complex and in general has an undesirable effect on the semiconductor. Take gold for example, which has a valence of one and leaves the bonding deficient of three electrons. Consequently, it will act as a triple acceptor and can be represented on an energy diagram as three discrete levels as shown in Fig. 3.9. Impurity atoms of this kind do not fit into the silicon lattice as well as the simple donor atoms. As a result of the slight distortion that occurs, the energy levels produced generally require more energy to donate or accept electrons and therefore are located further from the band edges than the simple dopants as shown in Fig. 3.9. In general this type of impurity level is undesirable and a major problem in semiconductor technology is to remove all such impurities that accidentally find their way into the solid, otherwise their effect might mask those of the desired dopants. Particularly troublesome atoms are copper, carbon, and oxygen. In order to obtain truly intrinsic behaviour in a semiconductor, it is necessary to reduce the concentration of electrically active ions to a value below that of the intrinsic concentration $n_i$. This presents the materials technologist with very severe tasks. In silicon, for example, we require the net doping effect $N_A - N_D$ to be less than $10^{16}$ m$^{-3}$ (at room temperature), that is one impurity atom per $10^{12}$ host atoms! The growth of intrinsic gallium arsenide with its room temperature value of $n_i \simeq 10^{12}$ m$^{-3}$ is beyond the capability of current technology and the process of compensation described earlier must be used to obtain 'quasi-intrinsic' material.

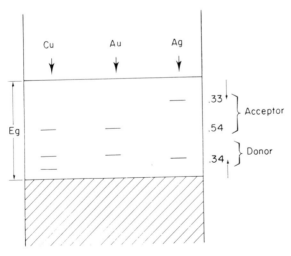

Fig. 3.9   A schematic diagram showing the deep energy levels of three 'unwanted' impurities, copper, gold, and silver, in silicon

In a similar vein, defects in the crystal lattice also produce deep levels. (A deep level is one sited well away from the band edges, in fact near to the centre of the forbidden gap.) A missing atom, for example (i.e. a vacancy), results in some of the bonds with the near neighbouring atoms being unsatisfied and such local regions will welcome electrons from the valence band. Such a defect would be acceptor like, thus producing holes (Fig. 3.10(a)). Alternatively, a defect such as an interstitial results in valence electrons being redundant from the normal bonding process and they can donate electrons to the conduction band and are donor-like (Fig. 3.10(b)). In practice, many stable defects that exist at room temperature may result in two or more forbidden levels per defect and therefore have a quite complex influence on the resulting semiconductor. In general such defects are unwanted and a great deal of effort is made to remove them. There are, however, certain circumstances where this is not the case. A prime example is the practical use of radiation damage in gallium arsenide. Crystalline defects introduced by ion bombardment produce deep levels in the band gap. These levels can trap and hold free electrons, thus converting n-type gallium arsenide into a compensated semiconductor (such defects also compensate p-type GaAs). Consequently, there are no free carriers and the material appears to be intrinsic. Because of the low intrinsic conductivity in GaAs, such regions are termed semi-insulating and may be used to isolate devices in a planar technology—see Chapter 7. Another use of deep levels is to provide recombination centres which shorten the lifetime $\tau$ of semiconductors. In silicon, both gold impurity levels and radiation damage levels have been used to decrease the lifetime and thus increase the switching speed of silicon thyristors.

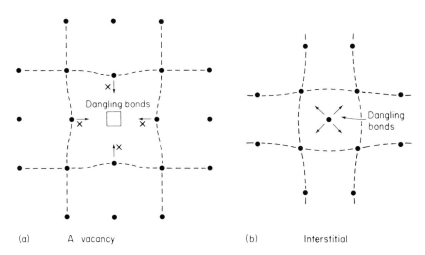

(a)        A  vacancy            (b)           Interstitial

Fig. 3.10    Illustrating the lattice relaxation around two simple defects: (a) vacancy; (b) interstitial

## 3.2  TYPE CONVERSION

Once a single crystal, which may be intrinsic, p-type or n-type, has been grown, the fabrication of a junction diode or transistor requires that some regions are made p-type and others n-type, with the impurity content very carefully controlled. If a section of crystal is p-type it can be made n-type by ensuring that the donor density $N_D$ exceeds the acceptor density $N_A$—this procedure is termed *type conversion* (i.e. we convert p-type to n-type or vice versa). One method of achieving this was described earlier in Chapter 2, namely to alter the dopant during the growth of a single crystal from the melt. The grown junction method was historically the first junction-type device to be fabricated, but is rarely used in current planar technology. It is, however, very valuable, for example, in making devices such as lasers.

Planar technology demands greater sophistication in the ways that type conversion is achieved. Both p- and n-regions with localized three-dimensional complexity are required and this may be achieved by two techniques—thermal diffusion and ion implantation. Traditionally, silicon devices and integrated circuits have been fabricated by thermal diffusion and the large capital investment in this technology made by industry over the last twenty years will ensure a continued commitment to its use. A more recent technology is that of ion implantation. Ion implantation is now a highly developed technology and offers many significant advantages over diffusion, In the case of silicon ICs ion implantation has produced improvements which are significant enough to justify the necessary capital re-investment needed for a new production-line technology. In the case of gallium arsenide a diffusion technology is effectively non-existent and ion implantation is the only viable technology for IC fabrication. These considerations will be discussed in this chapter.

Consider now the problems of achieving type conversion in some localized region of semiconductor, (as illustrated in Fig. 7.2). In this example the base semiconductor is intrinsic silicon and the required dopant is n-type (for example phosphorus). The object is to take the phosphorus from some external source and to place it on sites in the crystal that were occupied by silicon (i.e. to place it on a *substitutional site*). If the dopant impurity does not sit on such a site its electrical effects may not be as desired. One example of an undesirable site is the interstitial position shown in Fig. 3.10(b). Another occurs if the dopant becomes caught up on some crystalline defect such as a dislocation, or attached to some undesirable impurity such as carbon or oxygen. Some of these situations are very complex and the resulting electrical effects are beyond the scope of the present introductory text.

A solid such as silicon is a very dense concentration of atoms and thus, in order to be able to move the dopant from the surface into the bulk, work has to be done. In other words, energy has to be supplied to the dopant atoms. In atomic diffusion this source of energy comes from the thermal energy of the diffusion furnace, whilst in ion implantation it arises from the kinetic energy supplied to the ion beam by a particle accelerator. We will now discuss these two processes in more detail.

## 3.3 ATOMIC DIFFUSION

To produce atomic diffusion two conditions are necessary. Firstly, the density of impurity atoms must be non-uniform, with a source of material at or near to the surface and secondly, thermal energy is required to enable the atoms to migrate into the solid. In order to be able to understand the nature of the diffusion and to quantify the process it is necessary to look in some detail at its microscopic nature.

In the simplest terms, there are two basic ways in which atoms may diffuse: interstitial diffusion (Fig. 3.11(a)) and substitional diffusion (Fig. 3.11(b)). (There are in fact more ways than this but we will limit our discussion to these two simple processes.) Consider the first of these processes, the atom is sitting on the interstitial site which is a point of minimum potential energy. The surrounding nearest neighbour atoms are forced slightly outwards to accommodate this interstitial (note that the interstitial may be a foreign atom or a self-interstitial i.e. silicon in silicon). This atom will, like its neighbours, possess kinetic vibrational energy. If at some point in time this atom has more than the average amount of energy, it may force its way between a neighbouring pair of atoms onto one of the next nearest interstitial sites. (The number of such sites in silicon is four; it is a geometrical factor which depends on the crystal structure). To make this jump the diffusing atom has to do work and we can show this as a simple oscillating potential energy curve in the lower half of Fig. 3.11(a). The maximum size of this potential hill is called the activation energy and denoted by $E_a$. In many respects the substitutional diffusion process is similar. As in the case of interstitial diffusion it can be described by an activation energy $E_a$. Here, however, a further complication arises. In order to change from one substitutional site to the next, a vacancy in the lattice must be in the correct neighbouring position.

In order to quantify the diffusion process, consider a simple model where the atoms are vibrating with a frequency $v$ at the base of a potential barrier of the form shown in Fig. 3.11 and that the atoms have a Boltzmann energy distribution. The probability that an atom has an energy in excess of $E_a$ is therefore proportional to $\exp(-E_a/kT)$. Since the atom is vibrating with a frequency $v$ it will make $v$ attempts per second to strike the potential hill and pass over it. Furthermore, since in silicon there are four nearest interstitial sites, then the frequency of jumping is

$$v_j = 4v \exp(-E_a/kT). \tag{3.10}$$

In order to obtain an order of magnitude estimate of the jump frequency consider the situation at room temperature (300 K) with a typical experimental

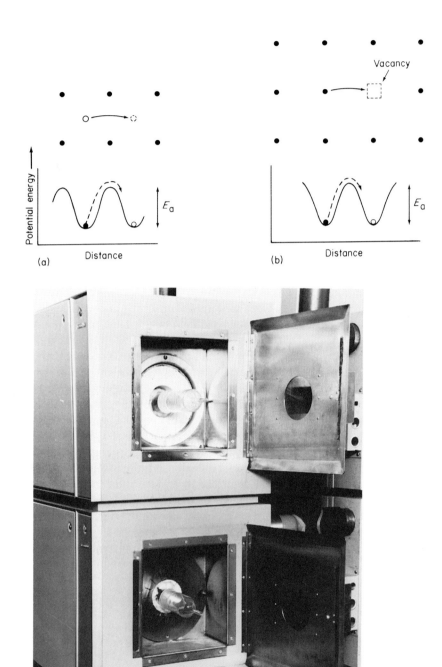

(a)

(b)

(c)

value of $E_a \simeq 1.0 \text{ eV}$ and $v \simeq 10^{13}/\text{second}$:

$$v_j = 4 \times 10^{13} \exp(-1.0/0.026) = 7.92 \times 10^{-4} \text{ s}^{-1},$$

i.e. an interstitial atom will jump approximately once every 1000 s and will rise as the temperature increases to a value typical of that required of diffusion ($\sim 1200 °C$).

In the case of substitutional diffusion the problem is similar in the sense that if a vacancy is located at a neighbouring lattice site the jump frequency is $v_j = 4v \exp(-E_a/kT)$. Here however, this factor has to be multiplied by the probability of finding a vacancy on that particular site. A vacancy is part of a Schottky defect and the probability of finding such a defect is simply proportional to the density of such defects in the crystal; i.e. $\exp(-E_s/kT)$, where $E_s$ is the energy required to form such a defect. Thus

$$v_j = 4v \exp\left(-\frac{(E_a + E_s)}{kT}\right), \tag{3.11}$$

where values of $(E_a + E_s)$ for substitutional impurities in silicon range from 4 to 5 eV and for self-diffusion in silicon $\sim 5.5$ eV. In this case the jump frequency at room temperature is

$$v_j = 4 \times 10^{13} \exp(-4/0.026) = 6.12 \times 10^{-54} \text{ s}^{-1},$$

i.e. a jump once every $10^{46}$ years!

The process described above is essentially a random walk. If, however, this mechanism is coupled with the existence of a concentration gradient of a specific atomic species, then a net migration will occur *down* the gradient (i.e. from the higher concentration to the lower). The laws governing this process are called Fick's laws and a simple derivation is provided in Appendix 4. There are two fundamental equations that may be derived. These are

$$\text{Fick's first law} \quad \phi = -D\left(\frac{\partial N}{\partial x}\right), \tag{3.12}$$

where $\phi$ is the net flux of diffusing atoms through a given surface located at $x$, $N$ is the concentration, $\partial N/\partial x$ is the concentration gradient, and $D$ is the diffusion coefficient defined in Appendix 4:

$$\text{Fick's second law} \quad \left(\frac{\partial N}{\partial t}\right) = D\left(\frac{\partial^2 N}{\partial x^2}\right). \tag{3.13}$$

This second differential equation is very important since its solution, subject to specific boundary conditions, enables diffusion profiles to be predicted and

Fig. 3.11 Illustrating the diffusion process and the accompanying potential energy versus distance curves: (a) interstitial; (b) substitutional diffusion; (c) part of a bank of impurity-diffusion furnaces

hence specific semiconductor device structures to be designed and fabricated. In practice there are two types of boundary condition that find general use. These will now be discussed in detail.


### 3.3.1  Constant source diffusion

This condition arises when a semiconductor wafer is subjected to a constant surface flux of diffusing atoms. In practice this occurs when a silicon wafer is placed in a gaseous environment at constant temperature. When this occurs the solid surface takes up a density of atoms appropriate to the solid solubility limit for that species at that given temperature. Some examples of the solid solubility limits are shown in Fig. 3.12. If the surface concentration $N_0$ of the diffusing atoms is held constant throughout the diffusion process, then the resulting profile becomes

$$N(x,\ t) = N_0 \operatorname{erfc}\left\{\frac{x}{\sqrt{(4Dt)}}\right\}$$

(erfc is the complementary error function), which, as indicated, is a function of distance $x$ and time as is shown schematically in Fig. 3.13. Two important points

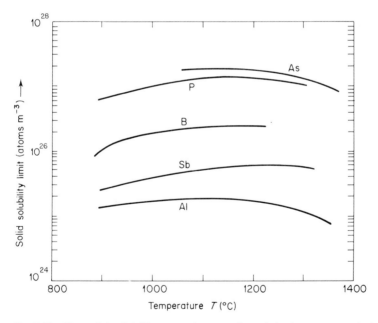

Fig. 3.12  The solid solubility versus temperature data for a range of atoms on silicon (after Colclaser (Ref. 3, Appendix 6))

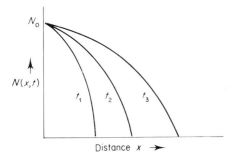

Fig. 3.13 A sequence of constant source diffusion profiles for three times $t_3 > t_2 > t_1$

to note are:

(a) The temperature dependence of $N$ appears in $N_0$ (which is a relatively slow dependence) and more importantly in the strong temperature dependence of the diffusion constant $D$.
(b) Diffusion does not give independent control over $N_0$ and the junction depth. This latter difficulty can be overcome by the use of ion implantation, as will be discussed later.

### 3.3.2 Instantaneous source diffusion

In this boundary condition the total quantity of diffusing atoms $Q$ is held a constant. Consequently, the total area under the diffusion curve (shaded area in Fig. 3.14) is a constant. The solution to Fick's second law becomes

$$N(x, t) = \frac{Q}{\sqrt{(\pi Dt)}} \exp\left(\frac{-x^2}{4Dt}\right).$$

The resulting sequence of profiles obtained at constant temperature but increasing time is shown in Fig. 3.14.

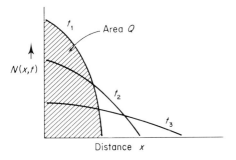

Fig. 3.14 A sequence of instantaneous source profiles for three times $t_3 > t_2 > t_1$

In practical device fabrication situations, it is frequently convenient to follow a sequence of diffusions consisting of a short constant source diffusion to place a given quantity of atoms $Q$ into the near surface region, followed by a longer instantaneous source diffusion to push the ions into the solid to produce the actual pn junction as is shown in Fig. 3.15. This is a very important procedure if a second pn junction is required, as for example in a pnp (or npn) bipolar transistor. In this way the surface concentration of the first diffusion can be reduced, thus allowing easy fabrication of the second pn junction.

Diffusion technology works well with semiconductors such as silicon or germanium. Its application demands only that the crystal can withstand the relatively high temperatures necessary for diffusion, without producing any adverse effects. Certain compound semiconductors are examples of solids where the diffusion process proves difficult. In GaAs, for example, long before the diffusion temperature is reached, the arsenic boils away from the surface region of the solid producing a gallium-rich surface unsuitable for device production. A second problem in some compound semiconductors—particularly those whose constituents come from groups II and IV in the periodic table (e.g. ZnS)—is autocompensation. Zinc sulphide is partly an ionic solid, thus the zinc and sulphur atoms carry electrostatic charges, the zinc positive and the sulphur negative. To preserve electrical neutrality these atoms must exist in the correct proportions. If, for example, a negative sulphur atom is removed from the solid, the localized region of the lattice is endowed with a net positive charge. To ensure that this does not happen, a positively charged zinc atom must be

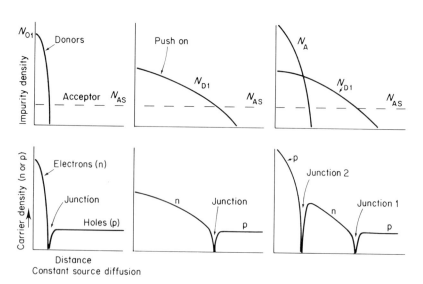

Fig. 3.15   Illustrating the impurity density and carrier density versus distance for the three diffusions needed to fabricate a pnp transistor structure

simultaneously rejected (i.e a zinc vacancy is introduced). This procedure of *automatic compensation* has been proposed to explain the difficulty experienced in producing type conversion in materials such as zinc sulphide and cadmium sulphide.

## 3.4 ION IMPLANTATION DOPING

In recent years a new technique called ion implantation has been developed as an alternative to diffusion for producing p- or n-type material. The technique has particular importance in instances when diffusion cannot be used or, indeed, where the technique offers significant advantages over diffusion. With ion implantation, dopant atoms are injected into the crystal not by the application of thermal energy, but by accelerating the atoms to a high velocity and implanting them into solid by virtue of their high kinetic energy. The essential feature of the technique is to take a beam of energetic particles emerging from an atom accelerator[†] and to direct them as a collimated beam onto the surface of the solid as is shown in Fig. 3.16(a). When such energetic ions enter a solid they are gradually brought to rest by a combination of multiple collisions with the target atoms and electrons in the solid. The collision with the electrons do not involve any significant transfer of momentum and hence do not cause the implanted ion to divert from its straight-line trajectory. It can in fact be thought of as a continuous frictional force very similar to a ball bearing moving in a viscous fluid. The collisions between implanted ions and target atoms on the other hand will produce deflections in the particle's trajectory as shown in Fig. 3.16. Furthermore, if the energetic ion communicates more than about 30 electron volts of energy to the target atom in the collision it can displace it, thus producing a vacancy and an interstitial (i.e. Schottky defect).

These are called radiation-induced defects and must subsequently be removed if the implanted ions are to dope the semiconductor correctly. In practice this involves heating the semiconductor to elevated temperatures in a clean and non-oxidizing environment. Heating the semiconductor to temperature greater than 600 °C does two things: it enables the radiation damage to anneal and also allows the implanted ions to move on to regular lattice sites of the host crystal as is shown in Fig. 3.17. Here one possibility is shown—an interstitial implanted atom finds its way onto a silicon vacancy. Returning to the actual implanted ions, these will be brought to rest at some distance $R$ (range) inside the solid. $R$ can be divided into its two components, the projected range $R_p$ and the transverse range $R_\perp$. Consequently, even for a monoenergetic ion beam, each of the implanted ions will have a different value of $R_p$ and hence the so-called range/profile will have a form similar to that shown in Fig. 3.16(b). This is in fact very close to being a gaussian curve with a maximum at the mean projected

[†]With a particle accelerator ionized atoms are accelerated to some desired velocity by use of an electrostatic potential.

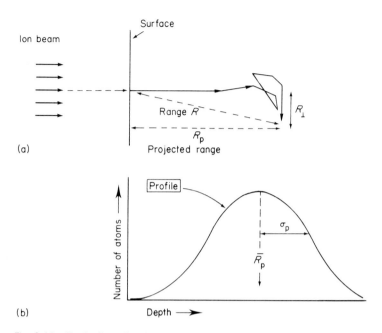

Fig. 3.16    Illustrating the ion implantation process: (a) the trajectory of a typical implanted ion; (b) the range profile resulting from a large number of ions

range $\bar{R}_p$ and a standard deviation (called the range straggling) of $\sigma_p$; both are strongly dependent on the mass of the implanted ion and on its initial velocity. The area under the curve depends on the total number of implanted ions. In Appendix 4 we provide some data on the projected range of some selected atoms in silicon and for GaAs as a function of the implanted ion's energy. Also shown is the range straggling $\sigma_p$ versus energy. This data is of crucial importance

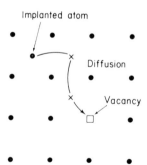

Fig. 3.17    Showing how an implanted ion sitting on an interstitial site moves to a vacancy by annealing and hence acts as a dopant atom

to the IC device designer. A second important point that must be stressed is that the shape of the implanted profile can be drastically altered if care is not taken to ensure that the incident ion beam is *not* lined up with one of the open channels that exist in single-crystal structures.[†] As an example of a typical ion implantation profile Fig. 3.18 shows the resulting three profiles for boron implanted into a silicon crystal at 30, 50, and 150 KeV.

A single implantation with a monoenergetic ion beam results in a single gaussian whose projected range $\bar{R}_p$ increases with increasing implant energy. This profile, by itself, is very limited for device fabrication since its rigid shape cannot be varied to satisfy some of the complex demands required by the device design engineer. Furthermore, the fall off in doping towards the surface can itself be a problem because of the difficulty of making the simple pn junction components, as is shown in Fig. 3.19. In this case the pnp structure would result.

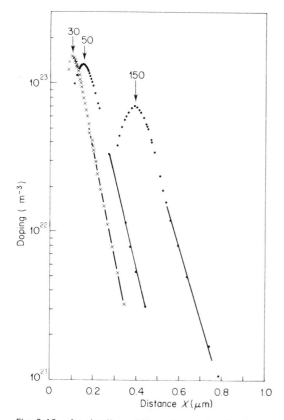

Fig. 3.18    A selection of three boron-implanted profiles in silicon

[†]This phenomenon, called *channelling*, is discussed in detail in Ref. 8.

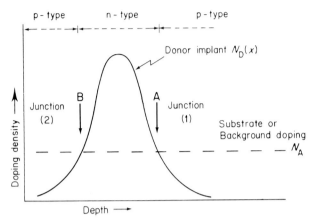

Fig. 3.19   A schematic diagram of a single n-type implant into a p-type substrate, resulting in a pnp structure

However, by altering the energy and dose of the implanted ions, the position of a pn junction (1) at A in Fig. 3.19 can be as carefully controlled as with diffusion. The implantation technique is more powerful than this simple procedure would suggest, since by conducting a series of implantations and carefully controlling the energy and dose of each implantation, the resulting profile can be tailored to almost any shape. Examples of this concept, Figs. 3.20 and 3.21, show two possibilities, obtained by summing individual gaussians. One example (Fig. 3.20) is a constant doping to some depth where the profile falls rapidly to zero. This could be used to fabricate a simple abrupt pn junction. The second example (Fig. 3.21) shows a profile that has large doping near to the surface but falls off to some constant doping level in the interior of the crystal. Such profiles are very

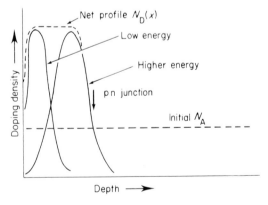

Fig. 3.20   Showing how two implants can be summed to give a flat abrupt pn junction

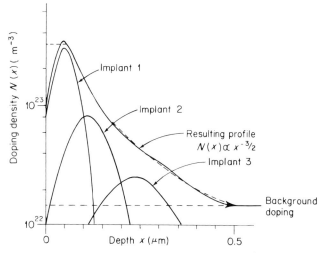

Fig. 3.21   Showing how three implants into n-type GaAs can be used to fabricate a profiled GaAs varactor

useful in producing variable-capacitance diodes (varactors) where the relationship between capacitance and applied voltage depends on the shape of the profile $N(x)$. In practice, it is usual to first of all define the shape of the desired (ideal) profile and to use a simple iterative procedure using a computer to determine the minimum number of implantations required to obtain that shape. Thus the engineer ends up with a specification of the depth ($\bar{R}_p$), and dose of ions required for a sequence of implantations. Using the projected range data in Appendix 4 the depth parameter can then be converted into an implant energy. In Fig. 3.21 we show a practical result of this process in fabricating a profiled Schottky barrier varactor device in gallium arsenide. A profile is required starting off at a doping of $3 \times 10^{23}$ atoms m$^{-3}$ at the surface and falling off as $N(x) = Bx^{-3/2}$, where $B$ is a constant, to a constant level of $1.5 \times 10^{22}$ atoms m$^{-3}$. In this case an approximation to the desired curve can be obtained by three implantations, i.e.:

| Implantation | Energy (keV) | Dose (m$^{-2}$) |
|---|---|---|
| 1 | 57 | $1.85 \times 10^{16}$ |
| 2 | 120 | $8.97 \times 10^{15}$ |
| 3 | 240 | $4.93 \times 10^{15}$ |

The summation procedure described above is of great importance, since it allows the design engineer to have control of the doping levels over the whole implanted profile, an important advantage over diffusion. A second advantage of ion implantation is its ability to obtain good uniformity over large wafer areas and good reproducibility from one run to the next. In order to obtain the good uniformity over the sample area, the ion beam is focussed to a spot and scanned electrostatically in a raster over the sample surface as is shown in Fig. 3.22. The excellent purity and cleanliness is achieved by a combination of the oil-free vacuum system and the use of the mass analysis magnet to ensure spectral purity of the implanted beam.

When fabricating complex device structures involving repeated implantations, as is required for integrated circuits, selective area doping is achieved by the use of masks in a manner similar to that used in diffusion. Here, however, it is frequently possible to use conventional photoresist for this purpose—this results in considerable simplification in the fabrication processing and simplicity is an excellent way of cutting costs. In this context, implantation offers further advantages over diffusion when working with sub-micron structures. The sideways scatter around a mask is $R_\perp \simeq \sigma_p$ and is much smaller than the corresponding sideways migration in diffusion. In the case of silicon technology these advantages are sufficiently significant to have justified the large capital investment required by the electronics industry to install and operate the ion implantation facilities which are necessary for commercial exploitation. When, however, we consider compound semiconductors such as gallium arsenide, ion implantation has yet another crucial advantage over diffusion. Gallium arsenide is unstable at high temperatures and surface decomposition occurs when the arsenic is lost. For this reason there is no diffusion technology available and the only alternative·for integrated circuit fabrication is ion implantation. These problems are being investigated at the present time with the view to producing monolithic gallium arsenide ICs for very high-frequency applications and for very fast logic circuits. In this context gallium arsenide is preferred to silicon because of its high electron mobility.

To end the discussion of ion implantation a brief description of a typical ion implantation facility will be presented. Figure 3.23(a) shows a schematic of an ion implanter.[†] In the ion source the atoms of the dopant species are ionized and a fine beam extracted and accelerated through some desired potential (up to a

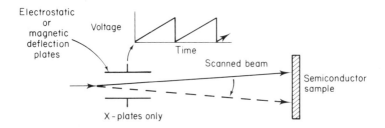

Fig. 3.22    Illustrating how a focussed ion beam can be scanned to achieve high uniformity over a crystal surface

[†]There are many variations of this particular system depending very much on the company that make the machine.

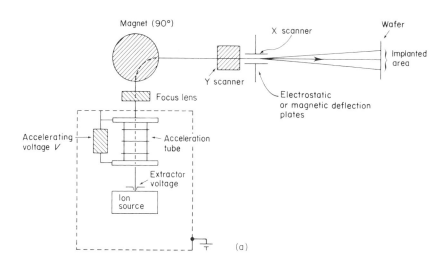

Fig. 3.23 (a) A schematic diagram of a simple ion-implantation facility; (b) a commercial ion implanter (permission of Whickham Ion Beam Systems)

few hundred keV). The accelerated ions are then passed through a magnet which has a field strength so arranged that only those ions with the mass of the desired species can pass through the exit aperture and on to the target chamber. The focussed beam then passes through the X-Y scanner which scans the ions in a raster over the sample wafer. The whole of the ion beam's trajectory occurs in a clean vacuum system and the target current provides the monitor on the total dose of ions implanted into the semiconductor. An example of a commercial ion implanter is shown in Fig. 3.23(b).

### 3.4.1  Example on diffusion

A silicon sample has a uniform donor doping level of $3 \times 10^{22}$ m$^{-3}$. It is desired to produce selective p-type doping of the near surface using boron diffusion. Further, it is required to keep the near surface doping density below that of the solid solubility limit of boron. Derive a simple design curve to relate the junction depth to the diffusion time for a drive-in diffusion at 1100 °C.

Since it is required to reduce the final surface acceptor concentration to less than solid solubility limit, the final diffusion process must be a drive-in using the instantaneous source diffusion equation:

$$N(x, t) = \frac{Q}{\sqrt{(\pi D t)}} \exp \left\{ \frac{-(x^2)}{4Dt} \right\}.$$

To do this we require to place $Q$ boron atoms into the near surface region. This can be done in two ways:

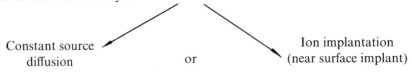

Constant source        or        Ion implantation
   diffusion                         (near surface implant)

Although ion implantation is the more versatile of these two techniques, in this example we will concentrate on diffusion.

#### Predeposition using diffusion

In this case diffusion is carried out using the constant source technique, i.e. an excess of the dopant is placed at the surface causing the surface to be doped up to the solid solubility limit $N_0$, where

$$N(x, t) = N_0 \, \text{erfc} \left\{ \frac{x}{\sqrt{(4Dt)}} \right\}.$$

Note, that in some applications where it is required to dope the near surface of the semiconductor p$^+$ (or n$^+$) this process can be carried out until the junction reaches the desired depth. Since for boron in silicon the solid solubility limit is reasonably constant above 1000 °C, the choice then is to select the temperature where the desired junction may be reached in a reasonable time (not too short (seconds) and not too long (many hours)).

### Decision: Temperature and time of predeposition diffusion?

Some experience is required here! Choosing a value of $T = 1000$ °C places $N_0$ in the constant region of the boron solubility curve (Fig. 3.12). The slightly low temperature ensures that the predeposition does not go too deep.

using $T = 1000$ °C $= 1273$ K and $1/T = 0.786 \times 10^{-3}$ K$^{-1}$,

we obtain, from Fig. A4.2,

$$D_1 = 2.7 \times 10^{-18} \text{ m}^2 \text{ s}^{-1} \text{ and } N_0 = 2 \times 10^{26} \text{ m}^{-3}.$$

The total number of atoms $Q$ deposited during predeposition is given by

$$Q = \left\{ \frac{N_0 (4D_1 t)^{1/2}}{\sqrt{\pi}} \right\} \qquad \text{Atoms m}^{-2}$$

$$\doteqdot 3.70 \times 10^{17} \sqrt{t} \qquad \text{Atoms m}^{-2}$$

| $t$(s) | 1 | 10 | 100 | $10^3$ | $2 \times 10^3$ |
|---|---|---|---|---|---|
| $Q \times 10^{-17}$ (m$^{-2}$) | 3.7 | 11.7 | 37.0 | 117 | 165.5 |

Times below a few hundred seconds are impractical and will be ignored. Similarly time above 10 000 s becomes expensive to implement. We shall try $t \simeq 2000$ s:

hence $Q = 165.5 \times 10^{17}$ atoms m$^{-2}$

and $\sqrt{(4D_1 t)} = 1.46 \times 10^{-7}$ m.

Substituting for $N_0$ yields $N(x)$ at $t = 2000$ s (Fig. 3.24). For comparison we note that with ion-implantation doping, the total number of atoms predeposited is simply

$$Q = \phi \text{ (the implantation fluence)}$$
and the peak concentration

$$N(x) = N_p = \frac{0.4Q}{\sigma_p}.$$

$\sigma_p$ for a particular ion/energy combination can be obtained from tabulated data (see Appendix 4).

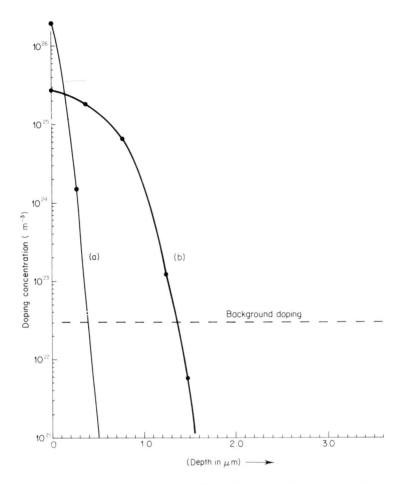

Fig. 3.24   Calculated diffusion profiles of boron in silicon (see text)

### Drive-in diffusion

Trial solution: 1100 °C for 60 min. (3600 s).

There are two conditions to consider. For high temperatures and/or long times the point where $N(x, t) = N_A$ (i.e. the junction depth $x_j$) goes deeper, whilst at the same time the surface concentration $Q/\sqrt{(\pi Dt)}$ falls. *Note that one problem with diffusion is that there is no independent control of these two parameters as there is with ion implantation.*

Substituting for $D_2$ (at $1100\,°C$) $= 3 \times 10^{-17}\ \text{m}^2\ \text{s}^{-1}$

and
$$t = 3600\ \text{s}$$
$$Q = 165.5 \times 10^{17}\ \text{atoms m}^{-2}$$
$$D_2 t = 1.17 \times 10^{-13}$$
$$(\pi D_2 t)^{1/2} = 6.07 \times 10^{-7}$$

hence
$$N(x) = 2.72 \times 10^{25}\ \exp\left\{-\frac{(x^2)}{4Dt}\right\}\ \text{m}^{-3}.$$

This is plotted as curve (b) in Fig. 3.24. The junction is defined as the point where $N(x_j) = N_D$ and from the graph is approximately 1.37 $\mu$m from the surface.

### Variation of junction depth with time

At the actual junction
$$N_D = N_A(x_j) = \frac{Q}{\sqrt{(\pi Dt)}}\ \exp\left(-\frac{(x_j^2)}{4Dt}\right).$$

Rearranging we obtain
$$x_j^2 = \ln\left\{\frac{Q}{N_D\sqrt{(\pi Dt)}}\right\}4Dt.$$

As a first-order approximation[†] we can ignore the dependence of $x_j$ on $\sqrt{t}$ within the logarithmic term and write
$$x_j = A\ (t)^{1/2}$$

where
$$A = 2.28 \times 10^{-8}\ \text{(using } t = 3600 \text{ within the log).}$$

As a check we can substitute $t = 3600$ into the equation and obtain
$$x_j = 1.37\ \mu\text{m in agreement with figure.}$$

This process can be repeated for different values of temperature and a whole range of $x_j(t)$ design curves obtained.

The other important design consideration is that relating to the surface dopant concentration $N(0)$
$$N(0) = Q/(\pi D_2 t)^{1/2}.$$

[†]This is a reasonable assumption since if $t$ varies from 1 to 10 000 then $\ln(\sqrt{t})$ varies from 1 to 4.6.

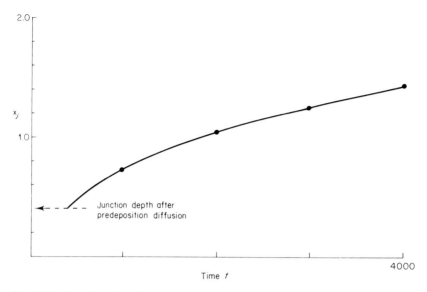

Fig. 3.25   Junction depth versus time for an 1100 °C drive in diffusion

Substituting for $Q$, and $D_2$

$$N(0) = 10.1 \times 10^{25}(t)^{-1/2}.$$

## Problems

3.1   Determine the value of $G$ (Equation 3.2) for intrinsic silicon at 300 K.

3.2   Determine the value of $G$ for intrinsic germanium at 300 K.

3.3   Determine the value of $G$ for intrinsic gallium arsenide at 300 K.

3.4   What is the band gap energy for an insulator?

3.5   Can $SiO_2$ be considered an insulator? Why?

3.6   A silicon wafer has an impurity doping density $N_D$ of $10^{22}$ m$^{-3}$ at 300 K. What is the resistivity of the wafer?

3.7   A silicon wafer has an impurity doping density $N_A$ of $10^{24}$ m$^{-3}$ at 300 K. What is the resistivity of the wafer?

3.8   Determine the mobility $\mu_n$ and $\mu_p$ for a silicon wafer with a total impurity concentration of $10^{24}$ m$^{-3}$.

3.9   Determine the mobility $\mu_n$ and $\mu_p$ for a silicon wafer with a total impurity concentration of $10^{22}$ m$^{-3}$.

3.10  Determine the conductivity for the silicon wafer in problem 3.8 with $N_D = 9 \times 10^{23}$ m$^{-3}$ and $N_A = 1 \times 10^{23}$ m$^{-3}$.

3.11  Determine the conductivity for the silicon wafer in problem 3.9 with $N_D = 9 \times 10^{21}$ m$^{-3}$ and $N_A = 1 \times 10^{21}$ m$^{-3}$.

# CHAPTER 4

# Insulating Films on Semiconductors

## Instructional Objectives

*This chapter describes the important role of insulating films in semiconductor device manufacturing and device operation. After reading this chapter you should be able to:*

a. Describe how oxides and nitrides are deposited on semiconductors.
b. Calculate thermal oxide thickness on silicon for given process conditions.
c. Explain how oxides and nitrides are used in semiconductor device manufacturing.

## Self-evaluation Questions

*Watch for the answer to these questions as you read this chapter. They will help point out the important ideas presented.*

a. Why are oxides used as masks in device manufacturing?
b. What is meant by the terms 'wet oxide' and 'dry oxide'?
c. What happens to donor and acceptor atoms in silicon during thermal oxidation?
d. When is $Si_3N_4$ sometimes used in place of an oxide?

## 4.1 INTRODUCTION

Insulating films play an important role in semiconductor device operation. They may serve to isolate one part of a circuit from another, or, as in the case of the field effect transistor, prevent direct current flow into the gate while allowing control over the current between source and drain. In GaAs a further important use of insulating films is observed. Arsenic is lost from the surface if the temperature exceeds 500–600 °C. An insulating film, usually $Si_3N_4$, is sometimes used to prevent this happening.

In silicon by far the most common and important film is $SiO_2$ and therefore most of this chapter will be devoted to a discussion of it. Some discussion of other types of film is given later in the chapter.

It was seen in Chapter 1 that an oxide film was used in the formation of our prototype integrated circuit. Its purpose there was twofold. First it was there to protect or passivate the junction between the p- and n-regions at the point where it came to the surface. The electric field is high in this region and without the protecting film the properties of the diode would be vulnerable to changes in humidity and various forms of surface contamination. A second purpose it served was to prevent the diffusion of the p-type impurity into the silicon everywhere except in the area selected for the position of the pn junction. In the final process of commercial ICs it is usual to deposit a protecting film over the whole circuit and remove it only over the bonding areas. This is known as *glassover* and serves to seal and protect the whole integrated circuit.

Thus we can summarize the uses of oxide film as follows:

1. Passivation of high-field regions on the semiconductor surface.
2. Masking or prevention of diffusion except in selected areas.
3. Masking for selective ion implantation.
4. As the insulating film in the gate region of MOS transistors.
5. Final circuit protection.

Uses 3 and 4 were not discussed in the context of our prototype IC and deserve futher mention.

An $SiO_2$ film can provide a very effective barrier to implanted ions so that if selected areas of the semiconductor are required to be ion implanted, then an oxide film of sufficient thickness may be used to prevent implantation in those areas where it is not required. Thus we could replace the diffusion step in the processing of our prototype by ion implantation. The p-region would then be formed only in the area required as before.

Formation of the gate region of an MOS transistor represents the most critical use of an $SiO_2$ film. The MOS transistor depends for its operation on a very high quality $SiO_2$ layer that separates the gate from the source channel and drain. This use will be discussed further in Chapter 7.

## 4.2  TYPES OF SURFACE OXIDES

In addition to the requirement for oxygen in the formation of $SiO_2$, silicon is obviously also necessary. Since the films may be formed on high-purity silicon wafers the latter may therefore be supplied by the wafer itself. Alternatively, the silicon may be supplied from some seperate source, usually in a gaseous form. In this case no consumption of substrate silicon occurs. This difference in the source of the silicon leads to the major classification of $SiO_2$ films.

## 4.2.1 Deposited oxides

A deposited oxide is formed when silicon and oxygen are both supplied from external sources. Under the right conditions of temperature, gas flow rates, and pressure, the two elements combine to form $SiO_2$ which is deposited on the surface. No substrate silicon is consumed in this process.

Referring to the uses of oxide films discussed earlier, final circuit protection is an example of the use of a deposited oxide. Clearly it would not be practical to consume silicon for the purpose of circuit protection since over a large part of the surface the film is not in direct contact with the substrate silicon. In addition, the other type of oxide, as we shall see below, requires temperatures of around 1000 °C for its formation and this would preclude its use in the final stages of processing.

The technique usually used for the formation of deposited oxides is known as chemical vapour deposition (CVD). It is a method used for forming many other thin layers, such as single crystal films (a form of epitaxy, see Chapter 2), polycrystalline silicon layers and films of silicon nitride (see Sections 4.9 and 4.8). Here we will concentrate on its use in the formation of $SiO_2$.

The silicon component of $SiO_2$ is supplied by silane ($SiH_4$) and the oxygen by pure $O_2$. A neutral carrier gas, usually nitrogen, is used. The substrate temperature range is 200–500 °C and the total gas pressure is usually close to atmospheric. Films are deposited at a typical rate of 70 nm per minute.

Recently, superior film coverage has been achieved by reducing the gas pressure to approximately 1 torr. However, the growth rate is lower at $\sim 10$ nm/minute. This technique is called low-pressure CVD (LPCVD).

## 4.2.2 Thermal oxides

If the silicon required to form $SiO_2$ is supplied by the substrate then the resulting film is a thermal oxide, so called because it has to be grown at high temperatures.

The immediate consequences of such a technique are (1) that a separate source of silicon is not needed, thereby simplifying the system. (2) The thickness of the substrate silicon gets less since silicon is being consumed. This is illustrated in Fig. 4.2. If the film thickness is $t$ then the substrate thickness is reduced by $0.45t$, whereas the overall thickness, including the film, is increased by $0.55t$.

In the growth of thermal oxides there are two basic means of supplying the necessary oxygen. In a 'dry' thermal oxide the oxygen is supplied as pure $O_2$ in gaseous form and the reaction is

$$Si + O_2 \rightarrow SiO_2.$$

If the oxygen is supplied in the form of steam (wet thermal oxide) the reaction is

$$Si + 2H_2O \rightarrow SiO_2 + 2H_2.$$

The growth rates for wet and dry oxides are quite different and will be discussed in Section 4.4.

## 4.3  PRACTICAL OXIDATION SYSTEMS

Examples of practical oxidation systems are illustrated in Fig. 4.1. In this figure two forms of CVD system are shown. The substrates are usually mounted on a

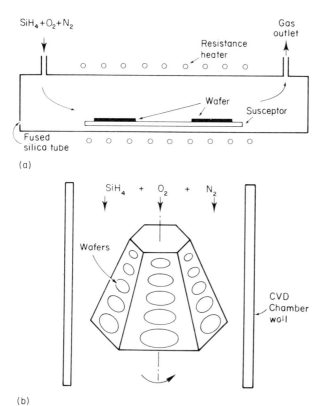

Fig. 4.1   CVD reactors for $SiO_2$ deposition: (a) horizontal system; (b) vertical barrel system for high throughput

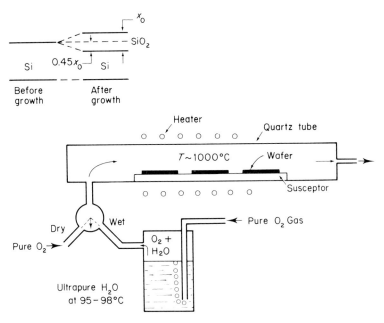

Fig. 4.2   Thermal oxidation system

graphite block or susceptor in a quartz tube, the whole being heated by a resistance heater. If uniform thickness films are required, then the substrate temperatures must be uniform over their whole surface. This demands very careful design of the heating system to give a high degree of temperature uniformity over the working length of the tube (the flat zone). An alternative high throughput production system is shown in Fig. 4.1. In this system the substrates are mounted on the faces of a barrel which is rotated slowly about a vertical axis in order to improve the temperature uniformity.

An example of a thermal oxidation system is illustrated in Fig. 4.2, which enables the growth of either wet or dry films by the turn of a control valve. An alternative means of producing high-purity $H_2O$ is to pyrolize $H_2$ and $O_2$. This produces water of very high purity but has the disadvantage of being more hazardous.

## 4.4   MECHANISMS AND DESIGN RULES FOR THERMAL SiO₂ FILMS

When the oxygen molecule and the silicon atoms from the substrate come together to form the $SiO_2$ molecule, there are, in principle, three ways in which this can occur, as illustrated in Fig. 4.3.

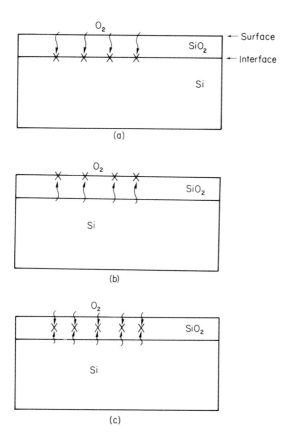

Fig. 4.3   Possible ways in which thermal $SiO_2$ films may be formed: (a) $O_2$ moves through film and $SiO_2$ formed at interface; (b) Si moves through film and $SiO_2$ formed at surface; (c) $O_2$ and Si move into film and $SiO_2$ formed within oxide layer

In Fig. 4.3(a), the oxygen molecule moves through the film and reacts with the silicon at the interface the film and the silicon. In Fig. 4.3(b), the silicon moves through the film and reacts with the oxygen at the film surface. Finally, in Fig. 4.3(c), both oxygen and silicon move into the film, the reaction then occurring somewhere within the film.

Radioactive tracer experiments have indicated that case (a) is really what occurs, the oxidant moving through the oxide film with the formation of the $SiO_2$ taking place at the interface (Fig. 4.3(a)).

We can envisage that the rate at which a film grows will be governed by two major factors:

(a) the rate at which the oxidant moves through the oxide film;
(b) the rate at which the chemical reaction to form $SiO_2$ occurs.

An analysis[†] taking account of both these factors (see Appendix 5) yields the following relationship between film thickness $x_0$ and time, $t$:

$$x_0 = \frac{A}{2}\left[\sqrt{\left(1 + \frac{t + \tau_0}{A^2/4B}\right)} - 1\right] \tag{4.1}$$

where $A$ and $B$ are constants for a given type of oxide at a given temperature, $\tau_0$ is a correction factor, necessary only in the case of dry oxides, given by

$$\tau_0 = \frac{x_i^2 + Ax_i}{B}, \tag{4.2}$$

where $x_i$ is an initial value of oxide thickness. It is seen from Table 4.1 that it is non-zero only for a dry oxide.

Equation (4.1) may be rearranged to give the growth time $t$ in terms of the thickness $x_0$:

$$t = \frac{A^2}{4B}\left[\left(\frac{2x_0}{A} + 1\right)^2 - 1\right] - \tau_0. \tag{4.3}$$

The constants $A$ and $B$ are extremely sensitive functions of temperature and are given by

$$A = K_1 e^{+E_1/kT}, \tag{4.4}$$

Table 4.1  Numerical values of parameters for oxide thickness calculations

| Parameter | (111) Silicon | | | (100) Silicon | | |
|---|---|---|---|---|---|---|
| | Wet | Pyrolytic | Dry | Wet | Pyrolytic | Dry |
| $K_1$ | 2.39 | 2.37 | 1.24 | 4.02 | 3.98 | 2.08 |
| ($\mu$m) | $\times 10^{-6}$ | $\times 10^{-6}$ | $\times 10^{-4}$ | $\times 10^{-6}$ | $\times 10^{-6}$ | $\times 10^{-4}$ |
| $K_2$ | 214 | 386 | 772 | 214 | 386 | 772 |
| ($\mu$m$^2$/hour) | | | | | | |
| $E_1$ | 1.29 | 1.27 | 0.77 | 1.29 | 1.27 | 0.77 |
| (eV) | | | | | | |
| $E_2$ | 0.71 | 0.78 | 1.23 | 0.71 | 0.78 | 1.23 |
| (eV) | | | | | | |
| $x_i$ | 0 | 0 | 0.02 | 0 | 0 | 0.02 |
| ($\mu$m) | | | | | | |

($k = 8.63 \times 10^{-5}$ eV K$^{-1}$).

[†] A. S. Grove, *Physics and Technology of Semiconductor Devices*, Chapter 2. Wiley (1967).

where $K_1$ and $E_1$ depend only on the type of oxidation employed (i.e. wet or dry), $k$ is Boltzmann's constant and $T$ the absolute temperature,

$$B = K_2 e^{-E_2/kT}, \tag{4.5}$$

$E_1$ is the activation energy of the diffusion process, and $E_1 + E_2$ is the activation energy of the reaction process.

The parameters $K_1$, $K_2$, $E_1$, $E_2$ and $\tau_0$ are given in Table 4.1 for both wet and dry oxidation. Use of these values in conjunction with equations (4.1) or (4.3) will allow the calculation of any oxide thickness to be made for a given temperature and time, or alternatively to predict the time for a given oxide thickness. Practical examples are given below to illustrate the calculations involved. However, first it is instructive to consider two limiting cases of equation (4.1).

For film thicknesses that are very small (i.e. during the early stages of growth) the oxidant moves through the film very readily and the limiting factor will be the rate at which $SiO_2$ molecules are formed at the interface.

The growth in this case is said to be *reaction-rate-limited*. If $t + \tau_0 \ll A^2/4B$, then equation (4.1) reduces to

$$x_0 = \frac{B}{A}(t + \tau_0), \tag{4.6}$$

where $B/A$ is known as the linear rate constant.

As the film thickness increases it becomes progressively more difficult for the oxidant to move through the film and the rate at which it can do so becomes the limiting factor. The growth rate is then *diffusion-rate-limited*. For $t + \tau_0 \gg A^2/4B$, equation (4.1) reduces to

$$x_0 = (Bt)^{1/2}. \tag{4.7}$$

The parameter $B$ is known as the *parabolic rate constant*.

The film thickness–time relationship is shown schematically in Fig. 4.4 where $x$ and $t$ are plotted on a logarithmic scale. The two regions corresponding in equations (4.6) and (4.7) are clearly seen.

### Example 1

A dry thermal oxide 0.2 $\mu$m ($0.2 \times 10^{-6}$ m) thick is required to be grown at 1100 °C on (111) orientation silicon. Calculate the time required.

### Solution

$$T(\text{in K}) = 1100 + 273 = 1373 \text{ K}.$$

Using the value of $k$ in Table 4.1, $kT = 0.119$ eV.

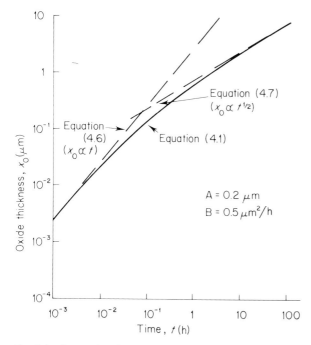

Fig. 4.4  Approximate and accurate forms for the dependence of oxide thickness on time

Using Table 4.1 and equations (4.2)–(4.5),

$$A = 1.24 \times 10^{-4} \exp (0.77/0.119) = 0.08 \ \mu m$$

$$B = 772 \exp (-1.23/0.119) \qquad = 0.025 \ \mu m^2/h$$

$$\tau_0 = (0.02^2 + A(0.02))/B \qquad = 0.082 \ h.$$

Inserting these values into equation (4.1) gives

$$t = 2 \text{ hours } 10.4 \text{ minutes.}$$

If the above example is repeated for a temperature of 800 °C, then 130 h oxidation time is necessary. This is clearly impractical, especially for a commercial process, and is one of the major reasons that typical oxidation temperatures are usually near 1000 °C.

*Example 2*

An oxide is grown for 130 min at 1100 °C on (100) silicon by passing oxygen through a 95 °C water bath. Calculate the resulting oxide thickness.

*Solution*

Again, using Table 4.1 and equations (4.2) and (4.5),

$$A = 4.02 \times 10^{-6} \exp{(1.29/0.119)} = 0.205 \ \mu m$$
$$B = 214 \exp{(-0.71/0.119)} \qquad = 0.549 \ \mu m^2/h$$
$$\tau_0 = 0 \ \text{for wet oxides.}$$

Substituting in equation (4.1),

$$x_0 = 1 \ \mu m.$$

A comparison of the results of the above examples shows that wet oxides grow much faster than dry oxides. An examination of the activation energies in the diffusion-limited region for wet and dry oxides shows that $E_2$ is nearly double that for a dry oxide. However, the pre-exponential factor $K_2$ is greater for a dry oxide, implying a higher growth rate.

Thus the lower growth rate for the dry oxide occurs because of the higher activation energy.

## 4.5   ADDITIONAL EFFECTS IN THE OXIDATION PROCESS

### 4.5.1   Enhanced growth at high doping levels

When thermal oxides are grown on heavily doped silicon the growth rates are increased above those discussed for more moderately doped silicon (see Section 4.4). Generally for both boron and phosphorus-doped silicon the growth-rate enhancement is not significant below $10^{26}$ per $m^3$ and above temperatures of 1000 °C.

Enhancement is greater for wet oxides as compared with dry, and films can be approximately four times thicker at 900 °C than for estimates based on the calculations of Section 4.4.

### 4.5.2   Effect of HCl

As will be discussed in Section 4.6, the quality of an oxide can be improved considerably if the oxidation is carried out with the addition of a few per cent of HCl to the oxygen when growing dry oxides. However, the growth rates are altered if HCl is used and the film thickness will be approximately one and a half times greater if 10% HCl is used.

### 4.5.3  Impurity redistribution

If silicon is consumed in the formation of $SiO_2$ during thermal oxidation, then the question arises of what happens to any n- or p-type impurities that are present in the silicon. Are they incorporated into the $SiO_2$ film along with the Si atoms, or are they pushed ahead of the moving Si–$SiO_2$ interface, thereby building up in the silicon?

The behaviour turns out to be different for different types of impurity and Fig. 4.5 illustrates the effect for the two major dopants, phosphorus and boron.

In the case of phosphorus a smaller amount is incorporated in the film than exists in the semiconductor, with the result that the remainder build up in the silicon close to the interface.

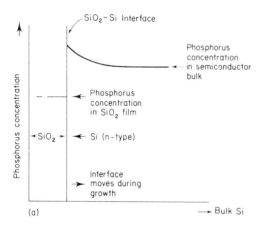

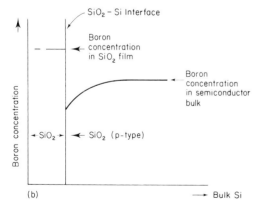

Fig. 4.5  Impurity distribution at oxide interface: (a) phosphorus impurity; (b) boron impurity

With boron it is different. A greater amount is incorporated in the film than exists in the semiconductor bulk, so that the silicon close to the interface becomes depleted of the impurity.

Clearly it is important in the design of any semiconductor device in which the impurity concentration in the semiconductor near the interface is important (an MOS transistor is a good example) that this effect is taken into account.

### 4.5.4   High-pressure oxidation

When the oxide growth is diffusion-rate limited, as is the case for thicker oxides, the rate at which the oxidation diffuses through the film can be increased simply by increasing the concentration gradient in the oxide film. This may be accomplished by increasing the impurity concentration at the surface, $N_s$ (see Appendix 5), and this increase may be achieved simply by increasing the partial pressure of the oxidant during growth. Thus an increase in growth rate can be achieved by carrying out the oxidation with the gas pressures at several atmospheres.

## 4.6   ASSESSMENT OF FILM QUALITY

The most critical application of a thermal oxide film is as the gate oxide in an MOS transistor. Thus the assessment of its quality is usually made in this context. Figure 4.6 illustrates the three important factors that can degrade the properties of an MOS transistor when present in its gate oxide.

First there is the fixed charge that exists at the interface of any thermally grown layer. It is independent of any potential applied to the transistor and is usually positive. Its magnitude is usually higher in a wet oxide than a dry, especially above 900 °C growth temperature. However, it can be reduced by up to an order of magnitude by subsequent annealing in nitrogen or argon. The interface state density may also be reduced if a small percentage of HCl is incorporated in the gas stream during dry oxide growth. However, this can affect the growth rates and should be taken into account (see Section 4.5).

An estimate of the magnitude of the interface state density may be made by measuring the capacitance of an MOS capacitor as a function of voltage. The presence of interface states causes a shift of the $C-V$ curve along the voltage axis, and by measuring this shift the estimate may be made.

The second type of defect that can occur in the oxide film is the presence of mobile ions which will move when a gate potential is applied over a period, and change the characteristics of the transistors. These are illustrated in Fig. 4.6. They are usually present because of sodium contamination and are nowadays less likely to occur in significant amounts. The presence and magnitude of these

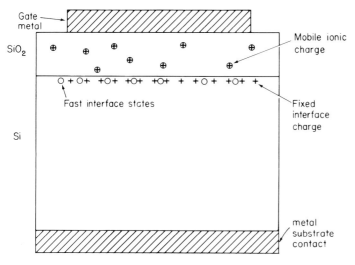

Fig. 4.6   Sources of charge in a practical MOS system

mobile charges may be assessed by making capacitance–voltage measurements before and after a high-temperature stress procedure in which the gate is held positive or negative while the device is held at a temperature of 200–250 °C. This causes the mobile ions to move to the side of the oxide that is at the negative potential, the high temperature speeding up this process. If there are mobile ions present in the oxide, this will show up as a shift along the voltage axis of the $C$–$V$ curve between the two measurements. The magnitude of the ionic charge can be inferred directly from the magnitude of the voltage shift.

The third type of defect is known as the fast interface state. Defects of this type occur near the oxide–semiconductor interface and are able to trap and release charge, depending on the bias conditions. The capacitance–voltage relationship in this case is shifted by varying amounts on the voltage axis. Thus the $C$–$V$ curve is distorted and by analysing the degree of distortion, the density of fast interface states may be calculated.

## 4.7  CHOICE OF FILM TYPE

The different types of film discussed have contrasting properties and it is therefore to be expected that different types are used depending on the application to which they are put.

It has already been argued that a deposited CVD film is more appropriate for final circuit protection or 'glassover'. The rest of this section will therefore be devoted to discussion of the relative applications of wet and dry thermal oxides.

The initial oxide in an MOS process is usually fairly thick ($\sim 1 \ \mu m$) and is usually grown as a wet oxide. The reasons for this are:

(1)   Quality requirements are not stringent.
(2)   High growth rates are desirable to form a film in a reasonable time period.

In contrast the gate oxide in an MOS process does have stringent quality requirements, and since the thickness required is typically 0.1 $\mu m$, the lower growth rates for a dry oxide still result in reasonable growth times. Thus it is usual for gate oxides to be grown using dry $O_2$.


## 4.8   NITRIDE FILMS

When considering the use of an $SiO_2$ film as a mask to prevent the diffusion of impurities into unwanted areas, it must be borne in mind that impurities will still diffuse to some extent in the $SiO_2$ film and what makes its use as a mask possible is the *relative* diffusion rates of any particular impurity in Si and $SiO_2$. The two major dopants, phosphorus and boron, diffuse relatively slowly in $SiO_2$, as compared with silicon and so $SiO_2$ will provide effective masking. Of course the film must be of adequate thickness, as discussed in Section 4.5. However, there are dopants which diffuse at a relatively high rate in $SiO_2$ so that the latter may not be used as a mask. Examples of such dopants are Ga, Al, Zn, Na and $O_2$.

It is found, however, that silicon nitride ($Si_3N_4$) is an effective mask for these impurities. Silicon nitride is usually *deposited* by the CVD technique so that no substrate silicon is consumed in its formation. The reaction between silane and ammonia is utilized (see Section 2.4). Substrate temperatures are usually between 800 and 1000 °C, a typical growth rate at 800 °C being 3 nm/min.

Nitride films give rise to a high density of interface states. It is therefore usual to grow first a thin $SiO_2$ film and then deposit the $Si_3N_4$. This system is used in a special type of MOS transistor known as the MNOS transistor.

The gate region consists of a thin grown layer of $SiO_2$ covered with a relatively thick layer of $Si_3N_4$. Defects which exist at the interface between the $SiO_2$ and $Si_3N_4$ can be charged or discharged by the application of fairly high potentials. The charge state can remain indefinitely under normal working voltages.

It is sometimes necessary to grow an $SiO_2$ layer over a region partly covered by a nitride film (see Fig. 4.7). In this case the $SiO_2$ growth over the nitride film takes its silicon from the latter, thereby converting it to $SiO_2$. For every micrometre of $Si_3N_4$ converted, 1.7 micrometres of $SiO_2$ is produced. However, this conversion rate is much slower than the $SiO_2$ growth rate on the silicon itself, so that for example during the conversion of, say, 0.1 $\mu m$ of $Si_3N_4$ in steam at 1100 °C (producing 0.1 $\mu m$ of $SiO_2$ over the nitride), more than 2 $\mu m$ of $SiO_2$ is grown over the silicon.

(a)

(b)

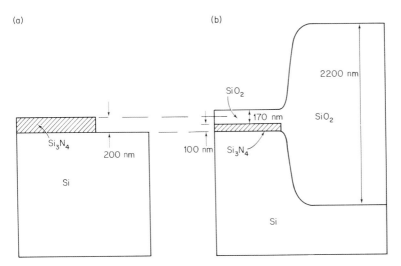

Fig. 4.7    SiO$_2$ growth over silicon nitride film (idealized profile): (a) before; (b) after

## 4.9  POLYSILICON FILMS

Although not strictly an insulator, the polycrystalline silicon (or polysilicon) film is finding increasing use in semiconductor technology so it is worthwhile briefly discussing its uses.

The polysilicon film differs from single crystal silicon in that it is composed of a matrix of single crystal grains, each having a random orientation to the others. The electrical conductivity of such films can be changed over a very large range (several orders of magnitude) by introducing donor or acceptor impurities into the films as is done for single crystal silicon. In contrast to the latter, however, the relationship between the number of donors added and the conductivity of the film is highly non-linear since many of the donors end up in the grain boundaries and are not electrically active.

One application for polysilicon films is as interconnects for MOS transistors (see Section 8.13). In this case the films need to have a low resistance so that in this case they are heavily doped. Another application is for passivating or protecting high-voltage integrated circuits. In this case the films need to have a high resistance and are therefore usually undoped. Sometimes they have oxygen incorporated in them to increase further their resistivity, in which case they are sometimes referred to semi-insulating polysilicon or SIPOS.

A final application is to the formation of highly efficient emitters for bipolar transistors. This enables shallower emitters to be fabricated and gives rise to higher-frequency operation.

## Problems

4.1  Why is nitrogen used as a neutral carrier gas with silane and oxygen to produce silicon dioxide?

4.2  If an oxide film is deposited at a rate of 70 nm/min, how long would it take to deposit 15 000 Å?

4.3  Given that a silicon wafer has a thickness of 0.4572 mm before oxidation, determine: (a) the thickness of the silicon after 1000 Å of $SiO_2$ is grown by thermal oxidation; (b) the total thickness of the silicon and the oxide after the 1000 Å oxide is grown.

4.4  How many millimetres of silicon is consumed to grow 15 000 Å of $SiO_2$?

4.5  Calculate the time required to grow a thermal oxide 0.1 $\mu$m (1000 Å) thick on (111) silicon at 1075 °C.

4.6  Calculate the time required to grow a dry oxide 0.1 $\mu$m (1000 Å) thick on (100) silicon at 1075 °C.

4.7  Calculate the time required to grow a wet oxide 0.1 $\mu$m (1000 Å) thick on (111) silicon at 1075 °C.

4.8  Explain the term 'pyrolytic oxide'.

4.9  An oxide is grown for 65 min at 1075 °C on (111) silicon by passing oxygen through a 95°C water bath. Calculate the resulting oxide thickness.

4.10  A dry thermal oxide is grown with oxygen and 10 per cent HCl. How much more oxide will be grown over that of an oxygen only system?

# 5

# Photolithography

## *Instructional Objectives*

*This chapter introduces the fabrication steps for a simple integrated circuit. After reading this chapter you will be able to:*

a. Describe a negative photoresist process sequence.
b. Describe the difference between negative and positive photoresist.
c. Describe photolithographic chromium and emulsion masks.

## *Self-evaluation Questions*

*Watch for the answers to these questions as you read this chapter. They will help point our the important ideas presented.*

a. When is photolithography used?
b. When are positive and negative photoresists used?
c. What problems occur when photoresist films are overexposed?
d. Why must the photoresist film have uniform thickness?
e. When are chromium masks needed?

## 5.1 INTRODUCTION

In Chapter 1 we saw the need for two pattern-forming stages or masking levels in the fabrication of our simple prototype integrated circuit. In a more complex

85

circuit more masking levels are usually necessary. However, in each case, in order to carry out batch processing the pattern for each individual circuit must be repeated over an array that is large enough to cover the whole wafer. We can identify several requirements that must be met by the technique used for forming the patterns.

1. It must be suitable for forming patterns in different types of surface film. In our example the first pattern was formed in $SiO_2$ while the second was formed in aluminium.
2. It is necessary to be able to 'align' each pattern level accurately with the preceding one.
3. The dimensional accuracy of each individual pattern must be sufficient to ensure proper alignment between the levels.
4. The chip size or repeat distance must be accurately maintained between different levels since any error would accumulate across the array and lead to misregistration on some circuits within the array.

The technique generally used in the semiconductor industry for all pattern formation is known as *lithography*. The most common one uses ultraviolet light and is called *photolithography*. Other types are *electron beam* and X-ray lithography, but these are not yet in common use. In order to illustrate how it operates we will use the example of Chapter 1 and show how the first level pattern in the $SiO_2$ (the opening for the $p^+$ diffusion) would be produced. As we proceed it will become apparent how the above requirements are satisfied.

The process has similarities with conventional photography in that there is a light-sensitive film, or emulsion, a controlled exposure to an image, and a development. A 'safe-light' environment in which to carry out the processing is also necessary to prevent over-exposure and loss of image. Figure 5.1 shows the various stages in the process.

(a) The silicon wafer is first oxidized (see Chapter 4).
(b) The wafer is coated with a U.V. light-sensitive film.
(c) A glass 'mask' containing the pattern for the appropriate level is brought into contact with the wafer. The mask can be used directly with its emulsion coating (an *emulsion* mask) or, if the pattern is transferred to a metal film, as a *metal* mask. The metal usually used is chromium.
(d) The whole is then exposed for the correct exposure time to a collimated beam of U.V. light, following which the mask if removed.
(e) The resist film on the wafer is then developed. At this stage the mask pattern has been reproduced in the photoresist, but as a negative.
(f) A suitable etch has to be used which will remove $SiO_2$ but will not attack the photoresist film or the underlying silicon significantly. When the $SiO_2$ in the uncovered areas and the photoresist has been removed, we have achieved the final objective of producing a pattern of 'holes' in the $SiO_2$ for the subsequent p-diffusion.

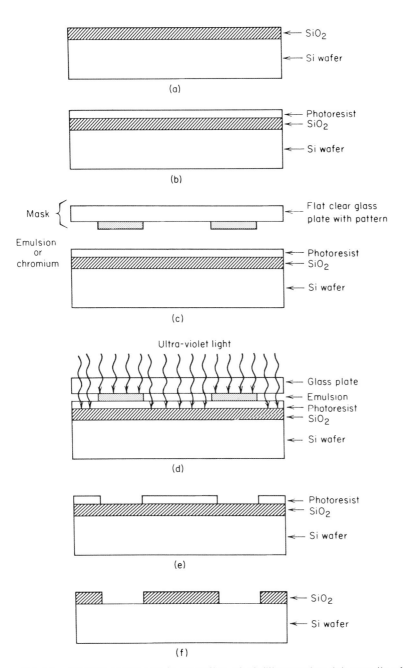

Fig. 5.1 Process sequence in negative photolithography: (a) growth of $SiO_2$ film; (b) coating with photoresist; (c) mask placed in proximity; (d) mask aligned and brought into contact; (e) photoresist developed; (f) oxide etch and photoresist removed

Now suppose we wished to produce the level 2 pattern in aluminium. The process would be generally similar but with the following modifications:

(a) The underlying film would be aluminium rather than $SiO_2$.
(b) The $SiO_2$ selective etch used in step (f) would be replaced by an aluminium-selective etch.
(c) A different mask would be used with the second level patterns on it.
(d) The second pattern would have to be aligned with the first so that the aluminium contact lies in the correct position over the p-region window.

The latter requirement is extremely important and will be discussed later in the section on mask alignment.

## 5.2  PHOTORESIST TYPES

Photoresists are supplied in the form of a liquid and consist of some U.V. sensitive substances in a solvent base. If a comparison is made between the mask and substrate in Fig. 5.1 (c) and (e) it will be observed that the pattern produced in the photoresist is the inverse or negative of that on the mask. For this reason those resists that are not removed on exposure to U.V. light and development are known as negative resists. Other types of resist which are *positive* working are removed on exposure and development. A comparison between positive and negative resists is made in Fig. 5.2, where the same final pattern in $SiO_2$ is produced by the alternative techniques.

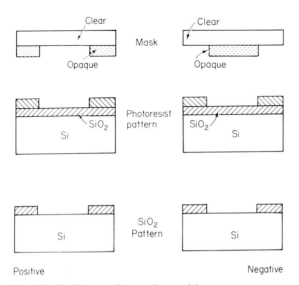

Fig. 5.2   Positive and negative resists

Negative resists consist of a base of synthetic compounds containing a few per cent of a light-sensitive agent which facilitates the formation of cross linkages between base molecules. This cross linking inhibits its dissolution so that exposed areas remain while unexposed areas are removed in an appropriate solvent.

Positive resists have a quite different chemistry, a light-sensitive compound in this case inhibiting dissolution unless the action of light has previously broken up the compound. In this case exposed areas are removed and unexposed areas remain.

If photoresist films are underexposed, there is a tendency for the pattern formation to be incomplete, and in the extreme a total loss of pattern can occur. With positive resist the film remains intact while for a negative resist the whole is removed.

If the film is overexposed, then windows opened in positive photoresist are slightly larger than the mask dimensions. This is because scattered light penetrates under the mask edges and exposes a small region of film not directly irradiated by the light source. Subsequent etching of the underlying $SiO_2$ film accentuates this enlargement, as shown in Fig. 5.3(a).

With negative resists the windows tend to be smaller than the mask dimension as illustrated in Fig. 5.3(b) but this is partly compensated for during the etching of the $SiO_2$.

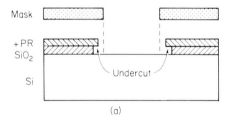

(a)

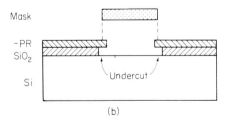

(b)

Fig. 5.3   Changes in window sizes with positive and negative resists: (a) positive resist; (b) negative resist

## 5.3    FILM THICKNESS

One of the important functions of the photoresist film is to protect underlying films of varying types (e.g. $SiO_2$, aluminium, polysilicon, silicon nitride). Thus it has to be of sufficient thickness to provide an effective barrier to the various etches to be used. Also, in the case of selective ion implantation it has to prevent ions from reaching the underlying silicon.

This sets a lower limit on film thickness. An upper limit is set by the need for good definition in the pattern as shown in Fig. 5.4. It is unlikely that sidewalls will be precisely vertical following development and clearly the thicker the film is, the greater will be the effect of any uncertainty as a percentage of the nominal window dimension. Thus the technique for coating wafer surfaces with photoresist must be capable of providing very uniform films with a high degree of control over the thickness.

The way in which thin, uniform films are obtained is illustrated in Fig. 5.5. Drops of liquid photoresist are introduced onto the slice held by a vacuum chuck which can be set to rotate accurately at some predetermined speed. The slice is then spun, centrifugal forces spreading the liquid outwards, any excess being thrown clear of the periphery. After a certain time the thickness stabilizes at a value dependent upon the rotational speed. The slice is then removed and baked in order to drive off the solvent and form a solid film which is thin and uniform over most of the slice. Some increase in thickness inevitably occurs at the edge of the slice due to surface tension in the liquid. Thus it may be necessary to discard later some circuits very close to the edge. Figure 5.5 also illustrates a commercial one-head spinner with accurate speed control, with very high acceleration and deceleration at the start of the spin cycle.

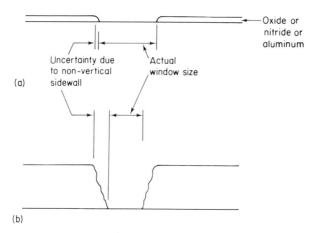

Fig. 5.4    Effect of non-vertical sidewall definition for (a) thin, (b) thick resist films

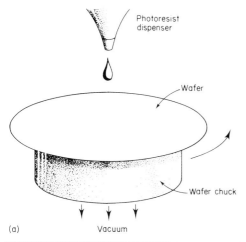

Fig. 5.5   Photoresist film formation: (a) coating with photoresist; (b) commercial photoresist spinner (courtesy Headway)

## 5.4   MASKS AND MASK MAKING

The prerequisite in the photolithographic process is a mask set each containing a uniformly repeated pattern appropriate to the particular processing level for which it is intended. A photograph of a typical mask is shown in Fig. 5.6 with an

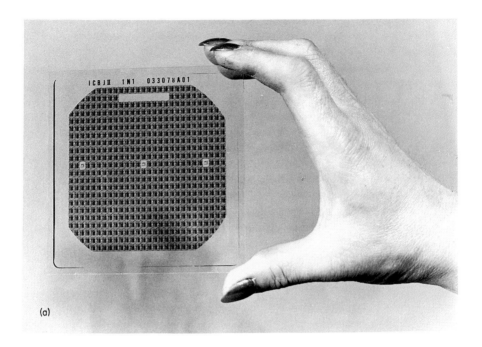

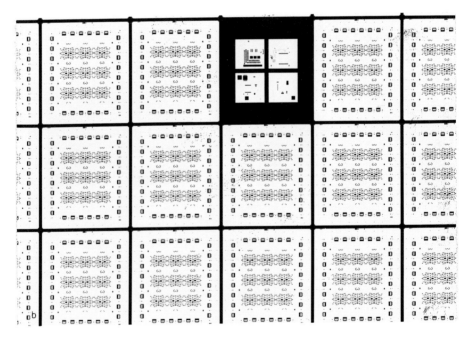

Fig. 5.6   Mask plate with enlarged view of one circuit: (a) mask photograph (courtesy Siliconix Ltd, Swansea); (b) enlarged pattern of above mask

enlarged view of some of the patterns also illustrated. The mask consists of an ultra-flat square glass plate, typically a few mm in thickness with one face covered with the pattern. It is, of course, this face that always has to be brought into contact with the coated slice during exposure.

The opaque regions are usually chromium and the patterns are formed by a similar photolithographic technique to the one discussed earlier. However, the additional complication present in mask making is that the pattern images have to be repeated in precise steps.

The way in which masks are produced is illustrated in Fig. 5.7. The integrated circuit designer has decided upon the number of masks needed for the complete process and has produced carefully dimensioned sketches of each mask. When the complexity of the integrated circuit is not very high the pattern is cut, say at a few hundred times actual size in double-skinned Mylar sheet. Mylar is a special plastic film which is dimensionally very stable.

One skin is clear and the other is usually red and therefore photographically opaque. The pattern is cut in the red film and peeled away where clear areas are

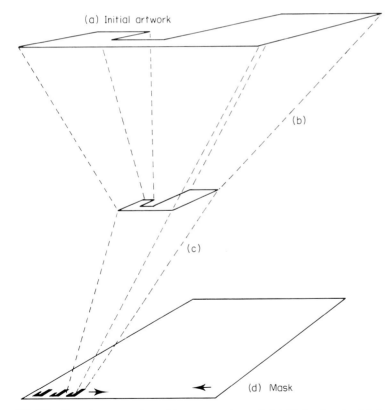

Fig. 5.7  Two-stage mask-making procedure: (a) initial artwork; (b) first reduction ($\sim \times 10$–40); (c) step and repeat ($\sim \times 10$); (d) mask

desired. The cutting is usually carried out on a precision table where the x and y positions of the cutter can be very accurately controlled. Only one pattern of the array is cut and the resulting sheet at this stage is called the initial artwork. The pattern is reduced by a factor of 10 or more using a reduction camera and the result is an intermediate photographic plate containing one pattern 10 times larger than finally required.

The final reduction is combined with precise stepping of the pattern in a piece of apparatus known as a step and repeat camera. This is illustrated schematically in Fig. 5.7.

An alternative mask-making technique which is increasingly used, particularly for high-density circuits, is for the designer to 'digitize' the required patterns on the screen of a special purpose computer. The patterns are then stored on tapes which are used to drive a mask-making machine directly. An obvious advantage of this approach is that standard designs may be stored and, if necessary, modified at will without having to write off any earlier effort and start afresh each time. Also it enables the designer to check his design with great accuracy since the facility to zoom in on any portion of the overall circuit and to view several levels simultaneously is readily available on such machines. Because of the emergence of powerful low cost workstations in recent years, the use of special purpose machines is progressively taking over, both for fully customised VLSI design, and for such semi-custom approaches as gate arrays and standard cell libraries.

## 5.5  MASK ALIGNMENT

As mentioned in Section 5.1, when more than one layer is to be used, each mask has to be precisely aligned with the previous layers. This is usually done by means of registration marks included on each mask. These masks are etched or patterned onto the silicon wafer along with the required pattern, for each mask used. The corresponding registration marks on the next mask to be used are then aligned under a microscope by means of a moveable stage. When alignment is correct the mask and wafer are brought into contact and the wafer exposed to U.V. light.

An alternative system, the project mask aligner, does not bring the mask and wafer into contact but projects one pattern at a time onto the wafer. This reduces damage to the mask and photoresist film but places stringent demands on lens quality and vibration control.

## 5.6  ELECTRON BEAM LITHOGRAPHY

Conventional photolithography using deep U.V. light (i.e. short wavelength) is at best limited to line definitions of just below 1 $\mu$m. A way of overcoming this problem is to use even shorter wavelength radiaition such as X-rays or electron beams. Although the potential of X-ray lithography is being evaluated it is not yet in common use and will not be discussed further. Electron beam lithography on the other hand has matured into a practical technology widely used by the specialist laboratories. The basis of the technique is to utilize the short wavelength of an energetic electron beam. For example, a 100 keV electron has a wavelength of $3.8 \times 10^{-3}$ nm and in practice the limit of line definition is related

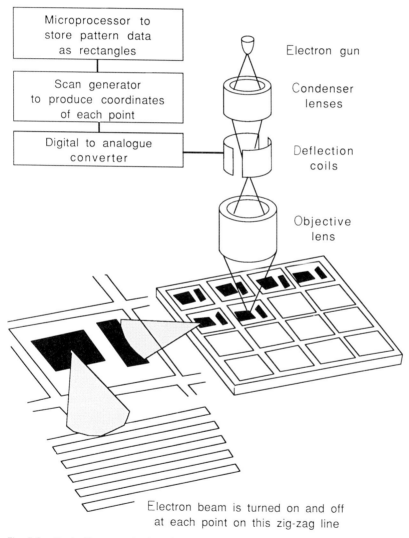

Fig. 5.8   Illustrating an electron beam lithography facility. Note the resemblance to a conventional scanning electron microscope (SEM) (Courtesy C. Wilkinson, Glasgow University)

to the electron beam optics rather than the wavelength as it is in optical lithography.

An electron beam lithography facility is essentially a modified scanning electron microscope. The electron beam is focused and can be deflected in two orthogonal directions as is shown in Figure 5.8. The desired pattern is written directly on to the chip by exposing the photoresist to the electron beam. To achieve this the beam is directed to a specific location and turned on. Automatic computer control moves the spot to scan out the desired pattern for exposure.

One of the practical difficulties with this technique is that it is relatively time consuming to carry out exposure over an IC wafer with immense detail. What can be done in practice is to mix the electron beam technique with conventional lithography. The bulk of the IC is defined with photolithgraphy and the fine detail (such as the gate of a MESFET) defined by using an electron beam. Figure 5.9 is an electron micrograph showing a sequence of 30 nm wide gold–palladium metal lines on 60 nm centre-to-centre spacing on a GaAs substrate and

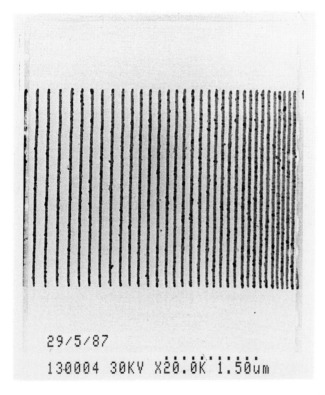

Fig. 5.9   30 nm wide gold–palladium metal lines on a 60 nm centre to centre spacing on a solid GaAs substrate. The Micrograph was taken using an SEM (Courtesy, C. Wilkinson, Glasgow University)

illustrates the potential of this technique. The technique is maturing rapidly, and offers great potential to future device design engineers.

## Problems

5.1  Why does photolithography use U.V. light?
5.2  What is meant by the term 'safe-light environment' when working with photoresist?
5.3  Describe what may happen if photoresist films are underdeveloped.
5.4  What materials are usually used to make photolithographic masks?
5.5  What can be used as an alternative to a glass mask when the complexity level of the circuit is large?
5.6  Why are electron beams sometimes used for lithography in place of light?

# 6

# Metallization, Interconnections and Packaging

## Instructional Objectives

*This chapter presents the techniques used to interconnect and package semiconductor devices. After reading this chapter you will be able to:*

a. Describe how metal interconnections are made in integrated circuits.
b. Describe how a semiconductor device is protected from chemical contamination.
c. Describe how semiconductor devices are mounted in packages.
d. Describe how wires are connected to integrated circuits.
e. Describe the role of the package in removing heat from the chip or device.

## Self-evaluation Questions

*Watch for the answers to these questions as you read the chapter. They will help point out the important ideas presented.*

a. Explain how the source material can be heated in a vacuum deposition system.
b. What advantage does electron beam evaporation have over resistance heating evaporation?
c. Why is electrolytic deposition inferior to vacuum deposition?
d. Why is $SiO_2$ deposited over an entire microcircuit?
e. Describe the difference between thermocompression and ultrasonic bonding.
f. Describe how internal heat generation is removed from the semiconductor.

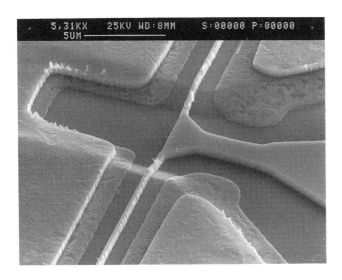

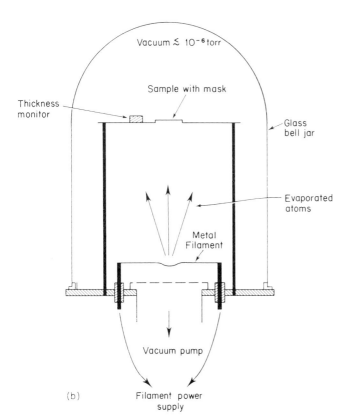

Fig. 6.1(a)  A scanning electron microscope picture of the metallization system for a MESFET transistor (Fig. 7.10). The source and drain regions are alloyed Au/Ge and the very fine gate wire is aluminium.

## 6.1  INTRODUCTION

The semiconductor device structure or integrated circuit is of no practical use until it is connected to the outside world. In practice this means that electrical terminals of high conductivity must be attached to the device to enable voltages to be placed across the device and current drawn from it. In most cases, this necessitates a good low-resistance ohmic contact to be made to the device terminals. (An ohmic contact is an ideal situation where the metal semiconductor interface has no intrinsic resistance). The majority of devices are fabricated by planar technology, where all the component devices, transistors, diodes, resistors, capacitors, and inductors are all embedded in the near surface region of the semiconductor. Consequently, all the metal contacts and interconnections must be deposited on the one surface. An example of a transistor metallization system is shown in Fig. 6.1(a).

## 6.2  THIN FILM DEPOSITION

With planar technology, the fabrication of contacts and interconnections and certain circuit elements by thin film deposition is very attractive. Vacuum deposition combined with photolithography can provide just the facility. There are two basic procedures for depositing thin films—vacuum evaporation, and sputtering—and we will now describe both of these in turn.

### 6.2.1  Vacuum evaporation

A schematic diagram of a vacuum evaporator is shown in Fig. 6.1(b). The material to be evaporated is placed in a small heater. The temperature of the heater is then raised until the source material vaporizes whence atoms are evaporated in all directions and condense on the inside surface of the bell jar and the substrate. By placing an evaporation mask in front of the substrate the geometry of the evaporated film can be controlled. The variations of this basic technique to suit the different materials are many. For example, the source material can be heated by resistance heating, or alternatively by electron bombardment heating.

With regard to the evaporation boats used in resistance heating there are many types now available commercially. Some examples are shown in Fig. 6.2. The most important criterion is that the boat metal must be one with a melting point which is very much higher than that of the metal being evaporated. Usual choices are tungsten or molybdenum (i.e. melting temperature of 3370 °C and 2620 °C respectively). Boat geometries can range from the simple coiled wire or indented ribbon to complex structures resembling small ovens. One very useful boat available commercially consists of a cone-shaped coil of wire, covered with

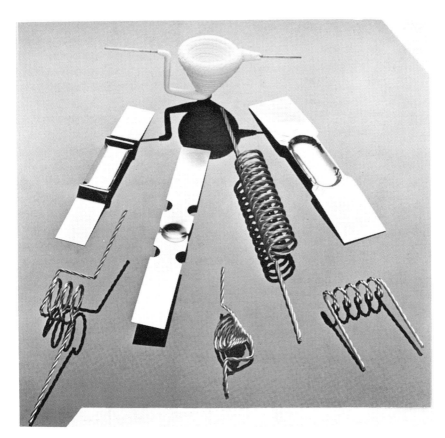

Fig. 6.2    Illustrating a range of evaporative filaments available commercially (permission of Nordico Ltd)

alumina coating (Fig. 6.2). This works particularly well with gold, silver, and the alloy gold–germanium.

Electron beam evaporation uses a focussed electron beam of high intensity to vaporize the source material. A schematic diagram showing the electron beam evaporation procedure is shown in Fig. 6.3. This technique is extremely powerful, partly on account of its potential purity and, equally important, the much wider range of materials that can be evaporated in this way. This is particularly important with the high-melting point metals such as platinum and tungsten, where resistance heating cannot be used. Another important feature of the electron beam technique is that the vaporization occurs only at a small spot in the source material. The supporting hearth is not directly heated and hence does not emit unwanted contaminant atoms as occurs in the conventional resistance-heated boats. Any foreign atom adsorbed on such boats is evaporated with the desired metal.

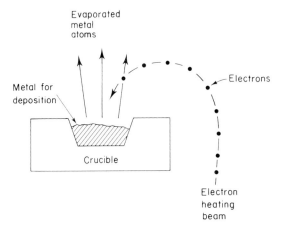

Fig. 6.3   A schematic diagram of an electron beam evaporation unit

## 6.2.2   Sputtering

The essential features of the sputtering technique are illustrated in Fig. 6.4(a). The bell jar is evacuated and an inert gas such as argon is bled through the needle valve to maintain a background pressure $\simeq 10^{-2}$ torr. The cathode is made of the material which it is desired to deposit on to the substrate. With the application of a high voltage (2–6 kV) between the electrodes, the inert gas is ionized and the positive ions accelerated to the cathode. On striking the cathode these will collide with the cathode atoms, giving them sufficient energy to be ejected. These sputtered atoms will travel through space, eventually coating the anode and, of course, the substrate.

The use of masking techniques can again be used to localize the deposition to desired areas. The simple sputtering system described above is a diode sputtering system. Practical coating units are in general more complex in detail (Fig 6.4(b)) but work on the same basic physical mechanism. Using the more complex sputtering systems it is possible to sputter annd deposit a range of metals as well as dielectric layers such as silicon dioxide.

## 6.3   PLATING

In comparison with the vacuum deposition techniques, plating is a much simpler and less expensive technology. Plating is essentially a process of electrolytic deposition of metal from solution and there are two basic variants: electrolytic and electroless. With electrolytic deposition, the surface on which the metal is to a be deposited is used as the cathode, and deposition occurs when the metal ions travel under the influence of the electrical field in the aqueous solution of metal salts. A schematic diagram illustrating this procedure is shown

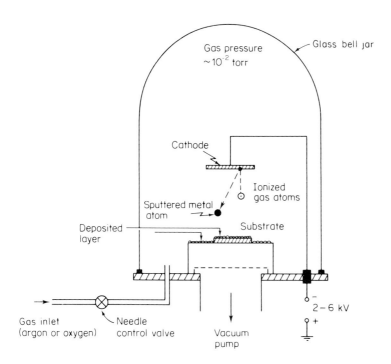

Gas pressure
~10⁻² torr

Glass bell jar

Cathode

Ionized
gas atoms

Sputtered metal
atom

Deposited
layer

Substrate

2 – 6 kV

Gas inlet
(argon or oxygen)

Needle
control valve

Vacuum
pump

(a)

(b)

Fig. 6.4   A diode sputtering unit: (a) schematic diagram; (b) a real system (permission of Nordico Ltd)

in Fig. 6.5. A d.c. source is connected between the substrate (cathode) and the metal to be deposited (anode). The deposition rate can be controlled by the current density in the bath. The electroless process is a technique for deposition through the catalytic action of the deposit itself, without the use of a source of external current. Electroless baths for the deposition of gold, copper, platinum, nickel, and palladium are available commercially.

In general terms, the plating process is inferior to those of vacuum deposition. High film purity is hard to achieve, the surface topography is relatively poor and thickness control harder to achieve. For these reasons plating is more generally used in a supporting role to vacuum film deposition. Hence the basic metallizations are deposited by vacuum deposition and plating is used to thicken up the metallization. (A good example is the production of a thick metal heat sink—a very important feature of high-power devices as is shown in the IMPATT diode of Fig. 6.6.)

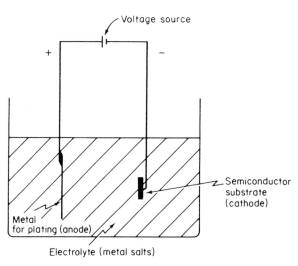

Fig. 6.5   A simple diagram illustrating the plating process

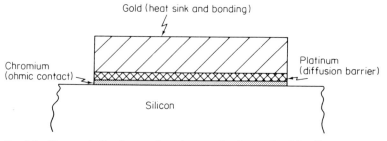

Fig. 6.6   The metallization system for an ohmic contact to silicon

## 6.4   METALLIZATION SYSTEM

For convenience the metallization systems used in microtechnology may be divided into two categories: (1) metals that perform specific electrical interaction with the semiconductor (such as ohmic or rectifying (Schottky) contacts) and (2) those used to interconnect the various circuit elements in an integrated circuit.

In group (1) the main constraint on choice is to ensure that the metal semiconductor contact has the correct electrical properties—ohmic or rectifying. In principle this can be achieved by matching the metal work function to that of the semiconductor.[†] In practice, however, the existence of a relatively high surface state density in both silicon and gallium arsenide precludes this possibility. Most of the noble and refractory metals result in Schottky barriers and rectification. To fabricate ohmic contacts, the semiconductor just beneath the metal must be doped degenerately (i.e. with a doping in excess of $10^{25}$ to $10^{26}$ atoms m$^{-3}$). Metal contacts to highly doped semiconductors always produce non-rectifying contacts because, even though barrier layers are produced, the depletion layers so formed are sufficiently thin to enable electrons to tunnel through.[‡]

One way of achieving such localized highly doped layers is by alloying certain metal contacts. This involves heating the metal–semiconductor system up to 300–600 °C in an inert atmosphere to allow the metal and semiconductor to intermix at the surface. In n-type GaAs, for example, ohmic contacts may be fabricated by depositing a mixture of 88% gold and 12% germanium as the contact metal, and when this is heated to 450 °C for 1 minute a good ohmic contact results (Fig. 6.7(a)). For silicon the near contact degeneracy is achieved by using ion implantation doping at low energies (Fig.. 6.7(b)) or by predeposition from a diffusion source. Such high surface doping levels cannot be attained in gallium arsenide and the thermal alloying process described earlier is necessary. For metallizations in general there are other demands made upon the choice which include: (a) good adherence to the semiconductor, (b) the metal layer and substrate must be capable of being selectively etched, (c) easy to bond wires to, and (d) does not readily degrade during operation at high current levels and elevated temperatures.

In this context problems can occur from thermally induced chemical reactions, interdiffusion or electromigration.

Thermally induced chemical reactions can sometimes occur between a metal and semiconductor, or indeed between two metal layers. A good example of this is between aluminium and gold. These two materials chemically react to form a purple compound, often referred to as a 'purple plague'. A second problem arises from interdiffusion at the boundary regions. The metal atoms diffuse into

[†] For full discussion of the physics of barriers see Ref. 6, p. 248, or Ref. 7.
[‡] Ref. 6, p. 248.

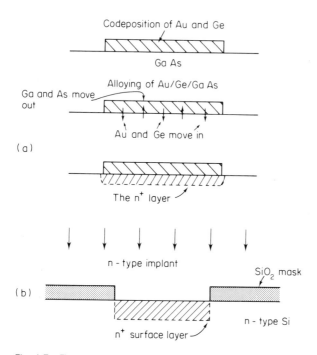

Fig. 6.7 The production of metal–n$^+$ ohmic contacts by (a) alloying Au/Ge to GaAs and (b) n$^+$ implantation into silicon

the semiconductor with a corresponding outdiffusion of the semiconductor atoms into the metal. Because of the power dissipated in semiconductor devices, the temperature rises above ambient and can enhance this interdiffusion. An example of a system which exhibits interdiffusion at relatively low temperatures (i.e. $\gtrsim 250\,°C$) is gold/GaAs and this phenomenon is a major cause of device degradation using this metal

The third problem identified above is electromigration. This is a result of the flow of a high density of electrons. When, for example, the current density in aluminium rises above $10^9\ A\ m^{-2}$ the momentum delivered by the electron flow to the atoms results in atoms gaining sufficient energy to migrate in the direction of current flow. The result of this is the formation of voids at one end of a metal stripe and the build up of globules at the other end. Eventually, this results in the metal forming an open circuit when the voids grow large enough.

In some devices no single metal can satisfy all these requirements and multi-layer metallizations have been developed. By way of an example, consider the high-powered silicon IMPATT diode shown in Fig. 6.6. Here the thin chromium layer is used as the ohmic contact with good adherence. The platinum is used as a stable barrier metal to stop the outer gold contact diffusing through the chromium into the silicon device, where it is known to cause rapid degradation at operating temperatures in excess of 300 °C. Gold is used for the upper metal partly because its ductile, oxide free nature allows easy bonding and because it is highly resistant to chemical attack and from oxidation during operation in oxygen. It must, however, be stressed that it is always desirable to use the simplest, least expensive metallization, and a compromise has to be made between performance and reliability on the one hand and system cost on the other.

## 6.5  SURFACE PROTECTION AND WAFER THINNING

At this stage in the processing the micro-circuit is a complete entity, ready to be interfaced with the external circuit. There are, however, three small but important steps to be carried out before we can proceed. First of all the whole of the active surface area needs some form of protection from physical damage (i.e. scratching) and from chemical contamination from its working environment. A good way of achieving both these objectives is to cover the whole of the wafer surface with a strong, impervious layer, such as silicon dioxide. This acts as a hermetic seal for all the devices and interconnection metallizations. Such layers can readily be formed by the sputtering process described earlier in this chapter. Silicon dioxide is not the only possibility; there has been considerable recent interest in silicon nitride as a passivation dielectric, particularly when working on gallium arsenide. An important point to note is that holes have to be cut above the metal contact pads which are needed for the external bond wires. This can be done by using photolithography.

The second problem arises from the finite thickness of the semiconductor wafer. When handling and working with the wafers they are subjected to stresses and have to be reasonably thick ($\gtrsim 150 \ \mu$m) to avoid breakage. For the final IC (or single devices), however, a thick layer of semiconductor lying between the near surface where all the action takes place and the metal surface to which it is to be attached is undesirable. The reason for this is that any heat generated during operation has to be extracted in this direction. A thin substrate enables a more efficient thermal conduction path to be achieved. Other considerations are:

(a) that the dicing process to be discussed in the next section is easier when dealing with thin wafers, and

(b) if diffusion has been used during processing, the back surface will contain unwanted diffused layers, which could interfere with its operation. These difficulties are removed by starting with thick layers and later lapping the rear surface, as shown in Fig. 6.8.

In cases where it is intended to mount the chip into a package using a eutectic process, it is desirable to cover the rear surface with a metal layer such as gold using a metal-deposition technique.

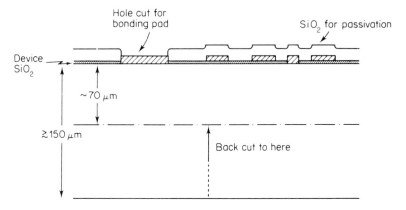

Fig. 6.8   Illustrating the back cut (thinning) and glass passivation of an IC

## 6.6   DICING, MOUNTING AND BONDING

The previous sections have been concerned with interconnecting a range of circuit elements situated in a planar configuration over the semiconductor. Metallization processes of this type are very cost effective, can be controlled with excellent precision, and consequently result in high reliability. A micro device or micro circuit produced in this way has now to be interconnected to much cruder and physically larger electronic components. Consequently, the device or integrated circuit has to be cut from the wafer, mounted in a suitable package and very fine bond wires must be used to connect the package electrical terminals to those on the chip.

In the first stage, the device of circuit chip has to be separated into its individual components—this may be a single device, such as an IMPATT diode for microwave work as discussed earlier or a self-contained integrated circuit containing many thousands of interconnected components (Fig. 6.9). This process, termed *wafer scribe*, can be achieved in many ways. Perhaps the most obvious is to use a diamond-impregnated circular blade to cut up the wafer (Fig. 6.10(a)). This technique is attractive since it readily lends itself to automation at the factory level: a whole bank of blades, mounted in parallel and correctly spaced, could simultaneously cut along the whole length of the wafer. One problem with sawing is the loss of material resulting from the finite width of the blade. This can be minimized by the use of thin saw blades.

A second method is that of cleaving the crystal along one of the major planes in the structure. When using the process the devices (or ICs) are laid down in such a way that their periphery squares are lined up with the known cleavage planes. Thus when the circuit is complete, a diamond-tipped scribing tool is drawn precisely across the wafer along the desired breakage line (Fig. 6.10(b)).

(a)

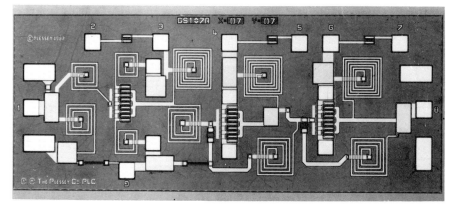

(b)

Fig. 6.9   (a) A digital circuit which operates at 5 Gigabytes/second. (b) GaAs microwave integrated circuit for operation at 3–6 GHz (both reproduced permission of Plessey Company plc)

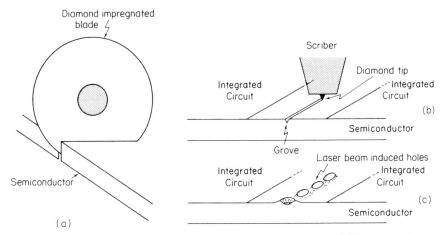

Fig. 6.10   Dicing of ICs: (a) diamond saw; (b) diamond scribe; (c) laser scribing

This is subsequently repeated over the whole wafer. To break the wafer, it is bent on both sides of the scribe line when cleavage occurs along the imperfection of the scribe line. As an alternative to scribing a laser may be used. A laser is pulsed to produce a sequence of small holes along the break mark, and the technique then works in a manner similar to the scribing technique with the holes functioning as the break defect (Fig. 6.10(c)). Because of the condensation of material in a crater around the hole (called *kerf*) this technique is best carried out along the rear surface where kerf will not damage the IC components.

Once the individual chips of semiconductor are separated by one of the processes described above they have to be mounted into some convenient package in order that the customer may handle and install such a device or IC in his circuit (if left in chip form it would require expensive equipment and specialist trained labour to use such integral circuits!). The chips may be mounted in packages by attaching them with either an epoxy resin, or preform, or eutectic 'die-attach'. The epoxy is simply a glue[†] which is placed between the rear surface of the chip and the package pedestal. Once these epoxy glues are baked to cure them they provide good adhesion.

In a similar manner preform bonding is achieved by placing a special piece of metal (called the preform) between the semiconductor and the package. This metal, when heated, melts at a low temperature and subsequently, when cool, sticks to both the semiconductor and the package. The third possibility involves an initial deposition of a thin layer of metal (such as gold on silicon) on the rear side of the chip. When heated to above the eutectic temperature ($\simeq 370$ °C for gold–silicon) the eutectic forms the required bond.

---

[†] Conductive or non-conductive varieties can be used, depending on whether the back surface is used as an electrical contact.

The final step in completing the device is to provide the fine wire bonds that go from the microscopic package terminals to the microscopic metal contact pads on the device (or IC) chip. This process is very labour intensive and demands skill, since it has to be carried out under a microscope. The concept is simple, a fine (25 $\mu$m diameter) wire has to be attached to the package and to the device as illustrated in the example in Fig. 6.11. This can be done by either thermocompression or ultrasonic bonding.

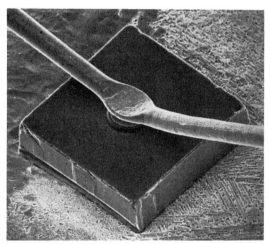

(a)

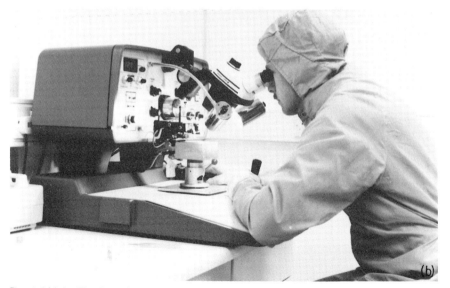

(b)

Fig. 6.11(a)   The bonding wire attachment to a low-power microwave-transferred electron device (reproduced permission of Plessey Company plc); (b) a commercial wire bonding facility.

The first of these techniques, as its name implies, uses a combination of heat and pressure to secure a firm mechanical and electrical contact between the two metal surfaces (wire and pad on the chip and wire and post on the package). A schematic diagram showing the basic steps in the bonding sequence is shown in Fig. 6.12. The end of the wire is melted so as to form a sphere under the influence of surface tension forces. The integrated circuit is heated to around 220–260 °C and the ball forced down onto the desired pad. Under the combined effect of the heat and the pressure the flattened metallic ball is forced into intimate contact with the pad and adheres to it as the tool is retracted as shown in Fig. 6.12(d). The tool is then taken over to the package contact and the same procedure repeated, but here there is no ball to compress, and instead the pressure is exerted on to a section of the wire as illustrated in Fig. 6.12(e). The heating is removed and the wire cut. An example of a device bond is shown in Fig. 6.11(a) which corresponds to a microwave-transferred electron device. The whole of the wire bonding procedure is carried out in special equipment, all the operations being observed through a microscope as is shown in Fig. 6.11(b).

The second bonding technique, ultrasonic bonding, is in many ways similar to thermocompression bonding in the sense that the two surfaces are again forced together by the use of a special bonding tool (Fig. 6.13). In this case, however, the package is not heated but the bonding is subjected to ultrasonic vibration in order to compress the surfaces together to achieve bonding. An important use of this technique is in situations where it is undesirable to heat the semiconductor chip. Figure 6.13 shows an example of a special bonding tool used for this work. Note how the bond wire is fed through a side hole to the base of the tool where the bond occurs. The usual wire used for ultrasonic bonding on aluminium pads is 99% aluminium and 1% silicon, but gold wire is reasonably successful on aluminium pads and very good when bonding to gold pads. These two bonding techniques can be combined in the so-called *thermosonic* bonding. In essence this is an ultrasonic bonder with a limited heating of the package ($\lesssim 150$ °C).

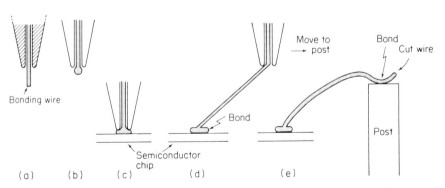

Fig. 6.12   A sequence of curves showing the thermocompression bonding process

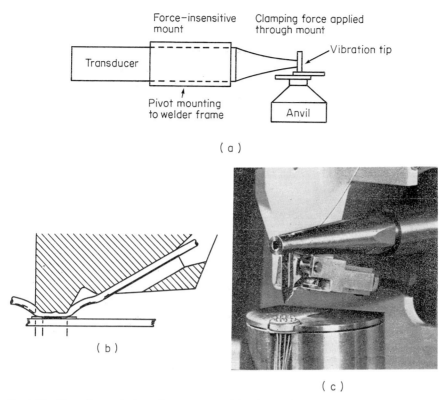

Fig. 6.13   The ultrasonic bonding process: (a) schematic diagram; (b) the bonding tip; (c) photograph of the tip

This is particularly valuable in situations where the chip cannot take the higher temperatures required for thermocompression bonding.

One very important point which emerges from the preceding discussion and is related to production costs is that the microtechnology processes divide themselves into two kinds. On the one hand we have the process steps that allow large numbers of identical ICs (or discrete devices) to be fabricated in the same sequence of process steps—a single wafer may contain many hundred circuits at one time. Furthermore they rely on proven 'clean' technologies which are readily automated. This is true of the metallization, surface passivation, ion implantation, and diffusion processes—indeed all the component technologies for the integrated circuit proper. From an industrial point of view the natural consequence of this is to work with larger-diameter wafers, thus increasing the number of given IC units per fabrication cycle (and hence per dollar!). With silicon processing, for example, the standard wafer diameter has increased from the two inch (5 cm) value of a few years ago to the present six inches (15 cm). The second class of processing involves such items as dicing and wafers, mounting in packaging, bonding, and testing. These processes are labour intensive because they are not amenable to batch processing. They are consequently relatively costly stages. This does, however, illustrate why an IC containing 1000 or more transistors plus a similar number of passive components is so much cheaper to produce than the equivalent discrete circuit.

### 6.6.1    Flip-chip and beam lead bonding

The wire bonding process discussed in the previous section is not always the best engineering solution to the problem and consequently specialist alternatives have been developed. The *flip-chip* technique, as its name suggests, is the procedure where a diode is mounted in a package in an upside-down fashion (i.e. with the device of IC chip substrate facing upwards). There are two reasons why this may be desirable. In some device structure such as microwave Transferred Electron Devices (TEDs) or IMPact Avalanche Transit Time devices (IMPATTs) the power density in the active region of the device can be as high as $10^{14}$ W m$^{-3}$. To remove this heat the diode must be placed on an efficient heat

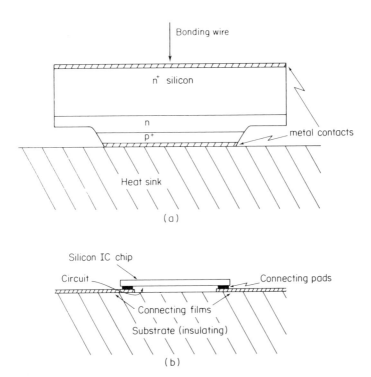

Fig. 6.14    Illustrating the flip chip process for (a) a single IMPATT device and (b) an IC chip

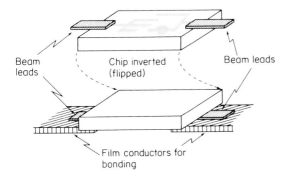

Fig. 6.15   Illustrating the beam lead process

sink. If an IMPATT diode is flip-chip mounted as shown in Fig. 6.14(a), the active region of the device is right next to the metal heat sink, thus removing the need for the heat to be conducted away through a relatively thick substrate which has poor thermal conductivity.

In the case of an integrated circuit the bond wires described earlier result in wasted space and parasitic stray inductances. Flip chip mounting of such chips can partly overcome these problems. Figure 6.14(b) shows how an IC chip is bonded to a substrate containing a pre-determined pattern of connecting wires on its surface. Note that raised bumps are attached to the chip, enabling it to make contacts only at specified points.

The beam lead bonding process is an alternative to flip chip and is very valuable in circumstances where stray parasitic elements must be minimized. This is very important at high-frequency operation as in, for example, micro-wave devices. The process is to form gold beam leads on to the chip while it is part of a larger silicon slice. Later the individul chips are etched out (i.e. not diced or sawn) leaving the 'beam leads' overhanging the edge of the chip as shown in Fig. 6.15. The chip is then inverted (flipped) and mounted onto a substrate in much the same way as the flip chip example described above.

## Problems

6.1   Explain the differences between vacuum deposition and electrolytic metallization techniques.

6.2   What is meant by the term sputtering'?

6.3   What is an ohmic contact?

6.4   What is a Schottky barrier contact?

6.5   Explain the gold–platinum–chromium ohmic contact system for silicon.

6.6   What happens if refractory metals are used to make contacts on silicon?

6.7   What type of metals are used to make ohmic contacts on GaAs?

6.8    What is electromigration and how does it degrade devices and ICs?

6.9    What is meant by the term 'passivation'?

6.10   Why is wafer thinning desirable for ICs?

6.11   Explain the advantages of the flip-chip bonding system.

6.12   Explain the disadvantages of the flip-chip bonding system.

6.13   Explain the advantages of the beam-lead bonding system.

6.14   Explain the disadvantages of the beam-lead bonding system.

6.15   Explain the meaning of package and device parasitics.

6.16   Explain the differences between a pn junction and a Schottky barrier.

# CHAPTER 7

# Fabrication of Devices and Circuit Components

## Instructional Objectives

*This chapter describes how basic circuit components may be fabricated. After reading this chapter you will be able to:*

a. Describe how Mesa diodes are made.
b. Describe a planar diode.
c. Describe planar transistor structures.
d. Describe a junction field effect transistor structure.
e. Describe a metal semiconductor field effect transistor structure.
f. Describe a metal-insulator-semiconductor structure.
g. Describe integrated resistors and capacitors.

## Self-evaluation Questions

*Watch for the answers to these questions as you read the chapter. They will help point out the important ideas presented.*

a. Explain the difference between a Mesa and a planar structure.
b. What is considered the optimum technology?
c. Discuss the importance of the buried layer in transistor operation.
d. Explain what 'field effect' means.
e. Explain the operation of charge-coupled devices.

## 7.1   INTRODUCTION

This chapter will link together the component technologies described in the preceding five chapters in order to illustrate how the different circuit components may be fabricated. To highlight the basic principles involved, the example discussed will be confined to simple components.[†] Extensions of these building blocks to complete integrated circuits (ICs) will be pursued in Chapter 8.

## 7.2   SIMPLE pn JUNCTIONS

By way of a starting example consider the fabrication of a simple pn junction diode. There are two possible ways to proceed, depending upon whether one is concerned with discrete components, or components in an integrated circuit (non-planar or planar technology).

### 7.2.1   Mesa etched diodes

Let us assume that we wish to fabricate a large number of discrete pn junction devices. A reasonable starting point is a slice of n-type silicon grown on a heavily doped $n^+$-substrate (Fig. 7.1). A heavily doped substrate enables good quality ohmic contacts to be made to the lower surface. The first step in the process is to produce a thin $p^+$-surface all over the sample. This may be achieved by a shallow boron diffusion or, as shown in Fig. 7.1, by a uniform sheet implantation of boron. At this point photolithography must be used to define the areas for the uppermost metal contacts. A uniform layer of photoresist is spun on the sample and baked in the manner discussed earlier. A photographic mask comprising circular dots is used to expose areas of photoresist to the U.V., which after developing produces the circular holes in the photoresist. The ohmic contact is then made to the $p^+$-region by vacuum deposition of a suitable metal (or alloy) over the whole surface of the slice. When the sample is now immersed in an acetone solution, the photoresist covering will be attacked and the metal overlying the resist will *lift off* leaving intact only the metal ohmic contacts made directly on the silicon. Success with lift-off depends critically on the existence of the weak region around the periphery of the hole in the photoresist.

In order to improve the diode quality it is sometimes desirable to mesa-etch to define and remove any damage (to be produced later by dicing) from the junction region. In order to do this photoresist is again spun on, once again using a photographic mask with circular dots, slightly larger than the first set,

---

[†] In order to clarify some of the features of device structures, the vertical scale will be enlarged with respect to the horizontal on the relevant diagrams.

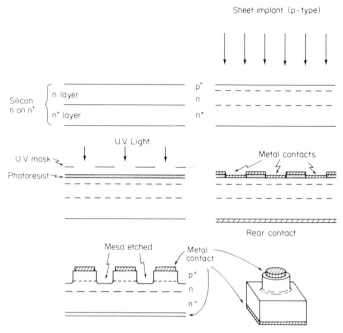

Fig. 7.1    Illustrating the stages in the production of discrete mesa pn diodes

but with the centres concentric with the original circular contacts; the regions outside these dots are exposed and removed by developing.[†] With the metal dots thus protected, the semiconductor may be etched away down through the pn junction to produce a Mesa, as is shown in Fig. 7.1. All that remains now is to coat the reverse $n^+$-layer with an ohmic contact, to scribe the slice and break it into its single (separate) components and to mount and bond these in a suitable package. The two ohmic contacts may themselves require an alloying process or a contact implantation to improve their quality. This depends very much on how heavily doped the initial $n^+$- and $p^+$-regions are.

### 7.2.2    Planar diode for monolithic circuits

As a second example we will consider a truly planar pn junction. The final device structure is shown in Fig. 7.2(a), which shows a pn junction formed by successive n- and p-type diffusions (or ion implantations) into a p-type substrate. The natural junction between the n-type diffused region and the substrate provides a good means of electrical isolation between the diode and any other device

[†] Alternatively we could have used negative photoresist with the reversed photographic images.

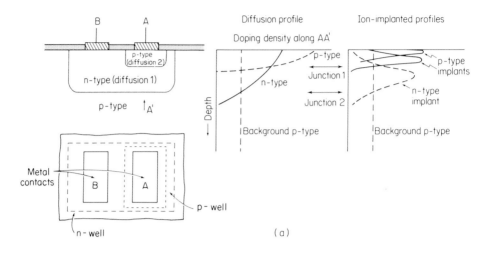

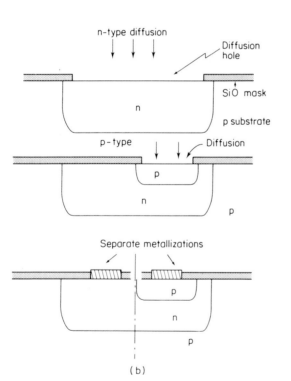

Fig. 7.2   (a) The structure of a planar pn junction diode together with the corresponding diffusion (or ion implantation) profiles; (b) the process steps for the planar diode

fabricated on the same substrate since, it must be remembered, in planar technology the diode might be just one component of an IC; Fig. 7.2(b) shows a sequence of events needed to produce any one of these diffused regions. (The photographic mask, expose, and develop steps are omitted for clarity.) This sequence of events is carried out firstly for the deeper n-type diffusion. Then it is repeated with a smaller hole (correctly aligned with respect to the first) for the second p diffusion. In each case a new $SiO_2$ layer must be formed in the manner described in Chapter 4. A third sequence of oxidation/photolithography/etching must be carried out to cut holes in the oxide at the appropriate places, in order to deposit the ohmic metal contacts.

At this juncture we must emphasize one important point. The basic processing steps and device geometry suggested above are not unique: many alternatives are possible. *In practice, the optimum technology is the one which minimizes the number of processing steps whilst at the same time guaranteeing a specified device performance.* In the case of a complex IC a trade-off is often required between these conflicting demands. For example, in the diode described above, the device quality could be improved considerably if shallow $p^+$ and $n^+$-regions were formed just below the metal contacts to ensure good ohmic behaviour. In silicon this could readily be achieved by ion implantation. However, this would increase the number of processing steps required, which of course would reflect itself in both processing time (i.e. cost) and in device yield. This latter point should not be ignored since the greater the number of processing steps, the greater the chance of a defect arising in processing and hence the lower the *yield* of good components. With regard to diode geometry, the situation is very dependent on the application envisaged for the diode. Figure 7.3(a) shows a linear device, but concentric circles would be just as easy, as is shown in Fig. 7.3(b).

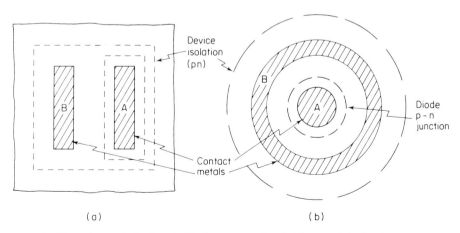

Fig. 7.3   Plan views of diodes of differing geometry: (a) linear; (b) circular

As was mentioned earlier a very important point relating to technology is that in any device/IC structure, there are inevitably different ways of arriving at the same end product, and the choice of which process to pursue is an engineering design exercise, taking into account parameters such as cost, yield, reliability of the end product, ease of automation as well as technical merit. Let us illustrate this with a simple example involving the device surface metallization aluminium. This could be used as device interconnections, as in silicon ICs, or alternatively as a Schottky barrier diode on a GaAs substrate for use as a rectifier or varactor diode. The principles are the same and in order to keep the argument simple we will illustrate the point with the latter example. The problem is very basic; it is required to deposit a circular aluminium pad onto the surface of an n-type GaAs chip. Figure 7.4 illustrates two different approaches to the problem. Consider first the left-hand branch. In this case the GaAs slice is covered completely with an aluminium film of the required thickness by vacuum deposition. Photoresist is then spun onto the aluminium surface and baked in the usual manner. At this

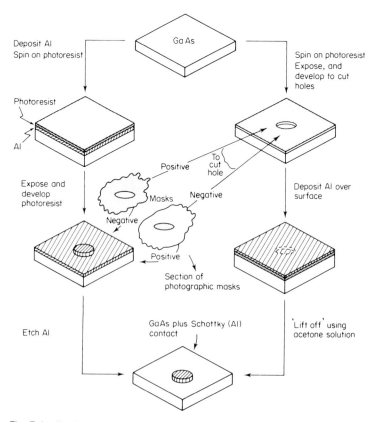

Fig. 7.4   Illustrating two ways of producing Al Schottky diode on GaAs

point, the process branches into two more alternatives, depending on whether one wishes to use positive or negative photoresist. For negative photoresist the photographic mask must be transparent in the area of the circular holes. The developing process then removes all the regions unexposed to ultraviolet light. With positive photoresist the reverse is true and a mask with circular black dots on a clear background is required. After development circular areas of aluminium are protected by photoresist. Consequently, when the slice is dipped in an aluminium etch all the aluminium except that protected by the photoresist will be removed, leaving the desired circular-geometry metallization.

The alternative, shown in Fig. 7.4, is the lift-off process described earlier. In this case the semiconductor slice is coated with photoresist and developed to produce a circular hole in the photoresist film. Here again there is a choice of positive or negative photoresist with their corresponding masks. The whole system is then coated with aluminium. Finally the photoresist is dissolved away in acetone solvent. The aluminium covering the photoresist breaks away (i.e. lifts off) leaving the required aluminium pads.

The processes described above are readily extended to complete IC metallization systems, where the geometry is more complex but the basic process steps are identical.

## 7.3  BIPOLAR TRANSISTORS

The planar diode process described in the preceding section is readily extended to produce bipolar transistor structures. These can be of two kinds, lateral and vertical. Cut-away diagrams of equivalent vertical npn and pnp transistors are shown in Fig. 7.5: in both cases they are fabricated on a p-type substrate. Consider first the npn transistor. It is very similar to the isolated pn diode described earlier, but here a third diffusion (or implantation) is carried out through a small hole and forms the $n^+$ emitter. A profile along the line AA' is as shown in Fig. 7.5(c). In order to achieve a low-resistance contact to the collector, the region under the collector terminal is doped $n^+$ as shown. The term 'vertical transistor' is appropriate because the major current flow from emitter to base to collector is as shown by the arrows in Fig. 7.5(a). With a p-type substrate the pnp transistor is somewhat simpler because the isolation junction is omitted (Figs. 7.5(b) and (d)). This implies one less diffusion. Note also that with minor modifications to the diffusion depths and the omission of the collector contact, this device is just the pn diode described earlier. This raises a very important point that is worthy of emphasis now. Since every transistor contains two pn junctions, transistors can readily be used as diodes, simply by using the same process steps but using different metallization geometries. This is illustrated in Fig. 7.6, which shows how the emitter–base junction can be used for rectification. In this case the base and collector are shorted together, thus ensuring that any emitter–base current that flows over to the collector will simply be added to

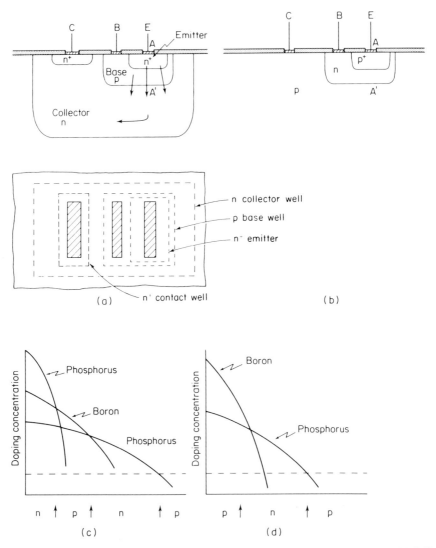

Fig. 7.5   Planar vertical transistor structures: (a) $n^+pn$; (b) $p^+np$; (c) and (d) the diffusion profiles required respectively

the base current. This allows standard diffusions to be used for both the regular transistors and the diodes. Differences are only required in the plan view geometry and the metallization masks. This drastically reduces the number of process steps needed. There are other possibilities for using the transistor processing to produce diodes. As an exercise, show how the base collector could be used efficiently as a diode.

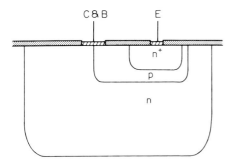

Fig. 7.6 Illustrating how the transistor structure can be used for diodes (in this example the collector–base junction is shorted)

An alternative configuration is the horizontal transistor shown in Fig. 7.7. Note here that the base and collector have changed positions and the current now passes horizontally from emitter to collector. This design appears somewhat simpler in structure but it might be remembered that the base width is now controlled by the photolithography masks and is in general limited to dimensions greater than $\simeq 1\ \mu$m.

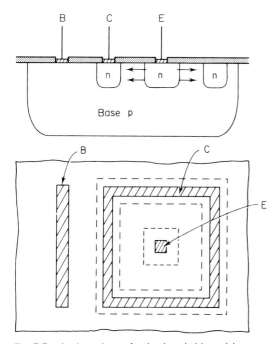

Fig. 7.7 A plan view of a horizontal transistor

### 7.3.1   Buried layers

For simplicity all the previous transistor diagrams have not included the buried layer now shown in Fig. 7.8. This is in essence an $n^+$ layer covering almost the whole active region and has the important effect of reducing the series resistance of the collector, whilst at the same time not degrading the collector–emitter breakdown voltage because there is no need to increase the doping in the bulk of the collector region. In essence, the current takes the shortest path to the high-conductivity buried layer, which shorts out the series resistance.

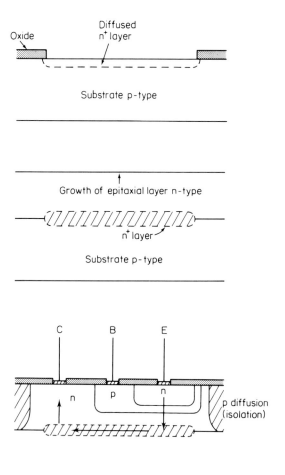

Fig. 7.8   The introduction of the buried $n^+$ layer into a transistor by epitaxial growth

## 7.4  JUNCTION FET (JFET)

The JFET structure is similar to a bipolar device as is illustrated by Fig. 7.9. The important part of the device is the n-channel formed between the lower isolation junction and the $p^+$-gate region. $n^+$-regions are again needed for the ohmic contacts to the source and drain. They also act to ensure that the parasitic resistances in series with the channel are reduced to a minimum.

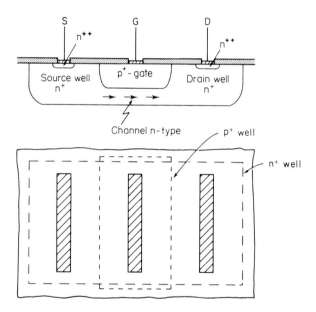

Fig. 7.9   A cross section of a simple JFET structure

## 7.5  THE METAL–SEMICONDUCTOR FET (MESFET)

Another variation on the basic JFET structure is the MESFET shown in Fig. 7.10. This is a JFET device in which the junction consists of a metal–semiconductor Schottky contact. It is particularly valuable in situations where pn junctions are not easy to form by diffusion (as for example in GaAs), and also in situations where the junction's parasitic capacitance must be kept small (a Schottky barrier does not exhibit diffusion capacitance).

A typical gallium arsenide device structure is shown in Fig. 7.10. The active device is constructed on an n-type expitaxial layer grown on a semi-insulating substrate. This is simply an intrinsic layer which, because of the wide band gap of gallium arsenide, has a low conductivity and is hence called semi-insulating.

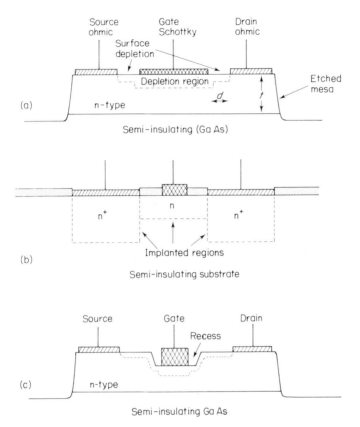

Fig. 7.10    A cross section of a MESFET structures in GaAs: (a) conventional mesa device; (b) ion implanted device; (c) recessed gate device

The thickness of the epitaxial layer controls the channel dimensions, since the electrons will not travel into the semi-insulating substrate. The source and drain contacts are ohmic and the gate is a Schottky diode.

### 7.5.1  Self-aligned gate MESFET transistors

The MESFET transistor shown in Fig. 7.10(a) is simplistic and for clarity the vertical dimension is greatly magnified. Of particular importance is the fact that the distance $d$ is considerably larger (1–2 $\mu$m) than the layer thickness $t$ ($\simeq 0.2$ $\mu$m). Furthermore a GaAs surface is really quite complex due to the existence of electronic surface states. These in effect cause a depletion layer to form at the surface which further reduces the effective thickness of the channel in the two regions between the source–gate and gate–drain contacts. Because of these effects the unwanted (i.e. parasitic) series resistance between the source contact and channel and the channel and drain contact increases. In consequence this degrades the transistors high-frequency performance. One way of reducing this resistance is to make the epitaxial layer thicker and to form a recess trench for the gate (Fig. 7.10(c)). The etching of this trench is undesirable in many applications as it takes us away from true planar technology.

An alternative process is shown in Fig. 7.11 and is a good example of the self-aligned gate process. Figure 7.11(a) and (b) illustrate the standard process of implanting the n-channel ($N_D \simeq 10^{23}$ m$^{-3}$) and the source and drain wells ($N_D \simeq 10^{25}$ m$^{-3}$). Note that in this second implantation the gate is made from a refractory metal and is used as a mask to define the second implantation. Tungsten silicide or titanium–tungsten silicide alloy have both proved valuable for such gate metals, the main difficulty being that the metal is left in place during the post implantation annealing up to temperatures of about 800 °C for 10 minutes. Gate metals such as aluminium or gold could not withstand such high temperatures.

Self-aligned MESFETs can be used for digital and analogue applications. The process is similar to that to be described for MOSFET fabrication and shown in Fig. 7.13.

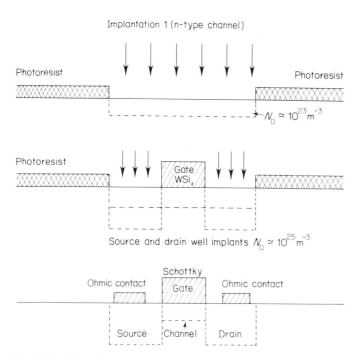

Fig. 7.11   Illustrating the self-aligned gate technology for a GaAs MESFET

## 7.6   METAL OXIDE–SEMICONDUCTOR DEVICES (MOS)

MOS transistors are in essence field effect transistors,[†] which utilize an MOS capacitor structure to control the channel region. Figure 7.12(b) shows the basic structure of a depletion mode n-channel MOSFET. It is very similar to the JFET shown in Fig. 7.9 (i.e., the $n^+$ source and drain regions). Here, however, a thin n-channel extends between source and drain, at the surface just beneath the silicon oxide layer. A gate metal is situated over the channel region which, with suitable biassing, can modulate the electrons in the channel. A positive gate bias will attract more electrons into the channel and push back the depletion region. Conversely, a negative bias will repel electrons and pull the depletion region towards the gate. Eventually, this will pinch off in much the same way as the JFET. One of the important advantages of MOSFETs over JFETs is the complete electrical isolation of the gate from the channel. The device shown in Fig. 7.12(a) is an enhancement mode n-channel MOST. The major difference here is that no in-built channel exists and in order to turn the device on a threshold positive voltage $V_T$ has to be applied to the gate. This is sometimes called the normally off or enhancement device and is very valuable in logic circuits since it does not consume power when it is not turned on.

   Ion implantation has been cited as the new arrival in the field of semiconductor technology. It is fortunate that the technique is ideally suited for planar technology. An excellent example of a situation where implantation provides an appreciable advantage over diffusion is in fabricating a high-frequency MOST. These devices are normally fabricated by diffusion (Fig. 7.13) where the source and the drain are introduced through windows in the oxide mask. Finally, the metal gate electrode is evaporated so as to span the gap between the source and the drain. Because of the limitation to the masking tolerance and the lateral spreading of the diffusing ions, the overlap in the gate and the source or drain can be comparable with the width of the gate. This gives rise to an inter-

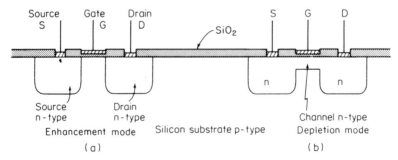

Fig. 7.12   Cross-sectional diagrams of n-channel enhancement and depletion mode MOSFETs: (a) enhancement; (b) depletion mode

[†] A full discussion of these devices is given in Ref. 6, p. 431.

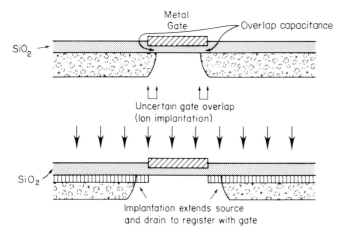

Fig. 7.13   Illustrating how ion implantation can be used to reduce the Miller capacitance of a MOSFET

electrode parasitic capacitance which limits its high-frequency behaviour. These capacitances can be particularly troublesome in integrated circuits with their high packing density. Figure 7.13 illustrates how ion implantation can reduce this capacitance effect. In this case the source and drain are produced in the usual way. This gate, however, is considerably smaller than the distance separating the drain from the source.

The complementary device—the p-channel MOST—is shown in Fig. 7.14. It is left as a problem for the student to understand how they may be operated. A combination of an n-channel and a p-channel MOST forms a very valuable unit in logic circuits, since when they are both turned on, the electron current transported by one device can be made to match and neutralize the hole current from the other. Thus no external current is required and such circuits consume low power. The technology is termed complementary MOS or just CMOS for short. The two device structures are compatible and can be run in parallel. CMOS is discussed further in Chapter 8.

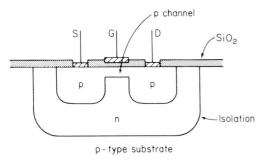

Fig. 7.14   A p-channel depletion MOSFET

### 7.6.1  VMOS field effect transistor

The VMOS FET uses the properties of anisotropic etches to produce this variant with its characteristic 'Vee' groove structure as shown in Fig. 8.4. It will be discussed in more detail in Chapter 8.

## 7.7  CHARGE-COUPLED DEVICES

Charge-coupled devices (CCDs) provide a new concept in devices since they use electric charge as their basic parameter rather than the conventional currents and voltage. A full discussion of their physics and uses is provided in Ref. 6 (Appendix 6). They are in essence delay lines where information in the form of a small package of charge can be injected into an input terminal and transferred in a controlled way along a basic sequence of electrodes. These devices have made dramatic advances since their discovery, in the late 1960s, primarily due to the fact that silicon was a suitable material for devices and they could thus benefit directly from the advanced state of planar silicon technology. A CCD consists of an array of basic cells joined in series. In its simplest form (a three-phase device), each cell comprises three MOS capacitor structures mounted parallel to each other, as shown in Fig. 7.15. By the correct sequence of bias on these three terminals, $\phi_1$, $\phi_2$, and $\phi_3$, a packet of charge in the left-hand MOS device can be moved in its entirety to the right into the first MOS device in the next cell. The

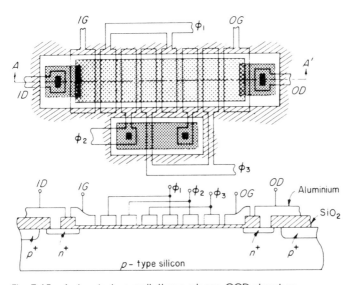

Fig. 7.15  A simple two-cell, three-phase CCD structure

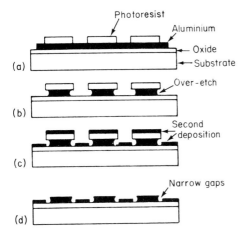

Fig. 7.16  The shadow etch technique

correct phasing and timing of these bias pulses by a clock causes the charge packet to move from one end of the device to the other. The simple structure shown in Fig. 7.15 consists of two cells in a buried channel CCD together with the input and output facilities (which will not concern us). The obvious compatibility with the standard silicon system is clear from this figure, i.e. the starting material is a p-type epitaxial layer and the active region for the charge transfer in an implanted n-type layer. In practical systems there are many ways of realizing this type of structure. Figure 7.16, for example, shows how the shadow-etch technique can be used to etch narrow gaps in a three-phase single-level aluminium gate structure of the type shown in Fig. 7.15.

## 7.8  PASSIVE CIRCUIT ELEMENTS—RESISTORS AND CAPACITORS

Earlier in this chapter it was shown how diodes could be obtained by modification to standard transistor structures; the same is true for some resistive and capacitive elements, as will now be discussed.

### 7.8.1  Resistors

One of the simplest structures is the diffused resistor shown in Fig. 7.17. In this example an n-type region is diffused or ion implanted into a background p-type semiconductor. This layer is electrically isolated by the pn junction and the silicon oxide and carries all the current between the two terminals. The resistance $R$ can be varied by the geometry (i.e. length $L$, breadth $s$, and

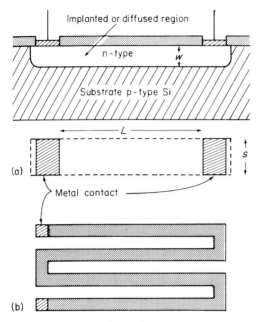

(a)

(b)

Fig. 7.17    The structure of a simple planar resistor

thickness $w$). A simple uniform profile can be obtained by the superposition of implants as was discussed in Chapter 3 (Fig. 3.21). For large resistor values the length of the resistor can be economically increased by meandering as shown in Fig. 7.17(b).

In this simple structure the control of $w$ is difficult if it is fixed by the epitaxial layer thickness ($\simeq 8\,\mu$) rather than diffusion or ion implantation. This can be overcome by the p resistor shown in Fig. 7.18—$w$ is now controlled by the base diffusion of the vertical BJT and for such a layer in silicon a sheet resistance of $\sim 150$ ohms/square is typical. As a result resistors of 50 ohms to 10 k$\Omega$ can readily be produced in this way.

Another possibility is to use essentially an FET structure (Fig. 7.9). A p-type

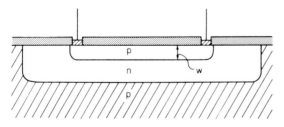

Fig. 7.18

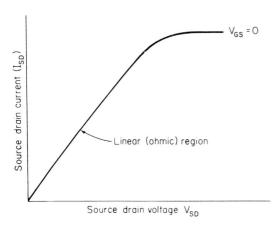

Fig. 7.19

diffusion limits the cross section of the resistor channel and enables much larger sheet resistances to be obtained and therefore higher resistances. The resistor in Fig. 7.9 does not have a bias on the gate. If this is metallized so that the gate can be biassed, then an electrically variable resistor can be produced. Linear (ohmic) behaviour is restricted, however, to the low-voltage regions well below saturation as shown in Fig. 7.19.

## 7.8.2 Capacitors

The small-signal models of transistors (bipolar and MOS) contain many capacitive elements and in principle any of these can be used as a capacitor in an IC. Figure 7.20(a) shows a typical capacitance voltage curve for a reverse biassed pn diode. The capacitance falls off with reverse bias—such devices are used to electronically tune electronic oscillators—they are called varactors. In a bipolar transistor there are two possible junctions that may be used: Fig. 7.20(b), for example, shows an emitter–base capacitor (i.e. an $n^+p$ diode), whilst Fig. 7.20(c) shows how a capacitor diode can be placed in the unused area of the diffusion stop. One of the problems with this type of capacitor is that the element will draw a finite leakage current in reverse bias (i.e. the reverse saturation leakage). In the terminology of the electronic engineer, we say that the quality factor (i.e. $Q$) of the diode is relatively low. The situation becomes much more difficult if the diode is taken into the forward bias, since in these circumstances the series resistance of the junction which is in parallel with the capacitance falls to a low value and the diode $Q$ will decrease drastically. Also, in the case of pn junctions the diffusion capacitance will increase rapidly in forward bias as shown in Fig. 7.20(a). A second difficulty is the variation of capacitance with voltage. This clearly cannot be used in situations where a constant capacitance is essential.

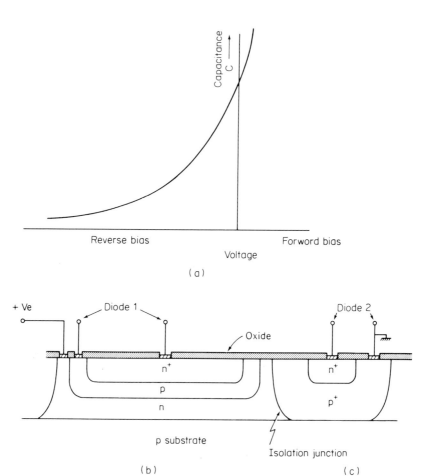

Fig. 7.20   (a) The capacitance–voltage curve for a pn junction; (b) and (c) two ways of realizing diodes from planar transistors

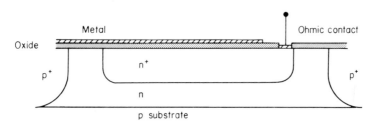

Fig. 7.21   A dielectric MOS capacitor

The difficulties identified above can be removed by using dielectric MOS capacitors. Components of this kind can be produced quite simply in silicon by using the surface $SiO_2$ already on the IC as the capacitor dielectric. It is required to place the dielectric between two highly conductive layers, and this can be achieved in the manner shown in Fig. 7.21. An $n^+$-layer is diffused or implanted into the epitaxial layer and contact made to it. An aluminium contact is laid down on the top of the silicon dioxide to complete the capacitor. The value of the capacitance is varied by a combination of oxide thickness and area.

## 7.9  GALLIUM ARSENIDE ICs

Silicon is the most important semiconductor in use at the present time. It is therefore perhaps correct that the IC technology described in the preceding sections has referred primarily to silicon devices. In principle all concepts can be transferred to gallium arsenide. However, gallium arsenide IC technology is in its infancy and it is becoming apparent from contemporary research and development that significant modifications are needed. One of the most important differences arises from the lack of a diffusion technology with GaAs. The problem is that the surface of GaAs cannot withstand the high temperatures needed for diffusion (the need for high temperatures was discussed in Chapter 4). Ion implantation is actively being pursued as a viable alternative and can claim some considerable success.

A second important difference is that GaAs docs not have a native oxide with the dielectric strength or the surface passivating qualities of the $SiO_2/Si$ system. At the present time no alternative deposited insulating layers such as $SiO_2$ or $Si_3N_4$ have been produced in such a way that the surface state density is sufficiently low to allow surface inversion to be obtained. Metal-insulator-semiconductor (MIS) devices comparable to the silicon MOS devices have not yet been developed. Active research is however being pursued in this direction. The only transistor structure as yet developed for serious GaAs systems is the MESFET device described earlier. This device is equally successful for discrete applications in microwave systems and as an IC element for analogue (microwave and millimetre applications) and digital ICs.

At the present time there are two substrate approaches being used:

(a) The use of n-type epitaxial layers on intrinsic (semi-insulating) substrates (Fig. 7.22). The intrinsic substrates act in the same way as the p-substrates in silicon. This only works in GaAs because of its wide bandgap. Sometimes a buffer semi-insulating layer is grown to protect the epitaxial layer from the diffusion of impurities from the bulk-grown compensated substrate. Device isolation can be achieved by etching, or by proton isolation.

(b) Direct implantation into semi-insulating substrates.

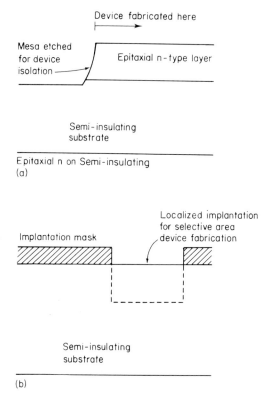

Fig. 7.22   Illustrating two approaches to GaAs IC fabrication: (a) mesa etched epitaxial layers and (b) direct implantation into semi-insulating substrates

In this case implantation occurs directly into the semi-insulating substrate with no epitaxial layer. The first implantation is used for the source and drain of the MESFET and with a re-mask cycle the channel is implanted.

As far as discrete microwave devices are concerned, the devices using epitaxial layers have produced the best performance. They are more nearly perfect than bulk-grown material, and consequently have a higher electron mobility and a higher cut-off frequency of operation.

The GaAs IC question is still to be fully resolved. However, at the time of writing of the second edition of this book (1989) the relatively simple microwave and millimetre (analogue) ICs are indicating clear success. Gallium arsenide foundries have been set up in Europe and the USA, and these can provide full services for IC design and fabrication of a wide range of high-frequency subsystems. The more complex digital ICs are experiencing some disappointing results partly on account of material problems giving low yield and partly because the speed of operation is not as fast as originally anticipated.

## 7.10  SPECIAL DEVICE STRUCTURES

The bulk of this chapter has dealt with the problems and solutions to IC technology. Simple discrete devices were considered as essentially a very simple IC and hence their fabrication followed similar technology steps. In some device structures special care is needed in bonding and packaging. For example, power devices need special heat sink considerations to prevent the device temperature from rising to an unacceptable level. Similarly, microwave devices require very special packages which have low parasitic inductive and capacitive elements associated with them. This is necessary because at the high frequencies of operation, greater than one gigahertz (GHz), the package properties would mask those of the device.

Current semiconductor technology is, however, much more diverse than this simple picture would suggest. There is a bewildering range of device structures which have been developed, some available commercially whilst others exist only in research laboratories. Consideration will now be given to an example which illustrates the direction in which one branch of the current microtechnology is moving.

### 7.10.1  The transferred electron device

The basic structure of a transferred electron device is essentially a thin region of n-type material (5–10 $\mu$m thick for operation in the region 8–18 GHz) with doping density $\simeq 5 \times 10^{21}$ atoms m$^{-3}$, sandwiched between two ohmic contacts (Fig. 7.23). Practical device structures need to be modified in the two following ways:

(a)  The active n-layer is grown epitaxially onto a relatively thick n$^+$-substrate (i.e. $N_D \gtrsim 5 \times 10^{24}$ m$^{-3}$). As in the case of the MESFET structure discussed earlier, it is usual to grow a thin epitaxial buffer layer first in order to inhibit the outward diffusion of unwanted impurities from the

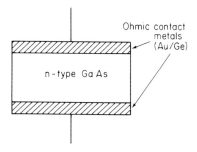

Fig. 7.23  A schematic diagram of a GaAs transferred electron device (TED)

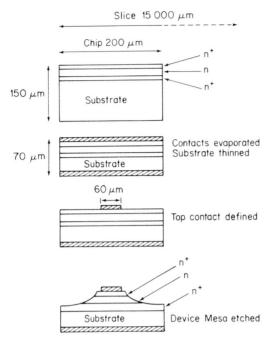

Fig. 7.24    The fabrication of a TED (reproduced permission of Plessey Company plc)

$n^+$-substrate into the active layer. This substrate achieves two goals:

(i)    It provides a rigid base on which to grow and process the active epitaxial layer;

(ii)    good ohmic contacts can be produced on $n^+$-layers.

(b)    A thin layer of $n^+$ is fabricated on top of the epitaxial n-layer. This enables good ohmic contacts to be produced at this surface. It is important to note, however, that in order to reduce the series contact resistances to an acceptable level both the metal contacts (gold–germanium) must be alloyed at around 450 °C for about 1 minute in order to increase the near contact ($n^+$) doping to a degenerate level (i.e. $N_D > 10^{26}$ m$^{-3}$).

## Problems

7.1    Describe a non-planar integrated circuit technology.

7.2    Describe a planar integrated circuit technology.

7.3    What are the advantages of horizontal transistor structures over vertical structures?

7.4    Draw a cross-sectional view of a vertical pnp bipolar transistor.

7.5   Draw a cross-sectional view of a horizontal pnp bipolar transistor.

7.6   Draw a top view of the transistor in Problem 7.4.

7.7   Draw a top view of the transistor in Problem 7.5.

7.8   Draw a cross-sectional view of a p-channel JFET.

7.9   Draw a cross-sectional view of a buried p-channel depletion-mode MOSFET.

7.10  What is the basic structure of a CCD?

7.11  An integrated circuit resistor is made of pure undoped silicon with a length of 2 mm and a cross-sectional area of 0.01 mm$^2$. What is the resistance of this resistor?

7.12  A resistor in an integrated circuit is made using the silicon substrate with a doping concentration $N_D$ of $10^{22}$ m$^{-3}$. If the length is 2 mm and the cross-sectional area is 0.01 mm$^2$, what is the resistance of this resistor?

7.13  Describe an MOS capacitor.

7.14  What are the advantages of GaAs over silicon?

7.15  What are the disadvantages of GaAs over silicon?

7.16  What does MESFET stand for?

# CHAPTER 8

# MOS and Bipolar Technologies and Their Applications

## Instructional Objectives

*This chapter describes how basic integrated circuits can be fabricated by using both MOS and bipolar techniques. After reading this chapter you will be able to:*

a. Explain bipolar technology families.
b. Explain MOS technology families.
c. Describe an integrated circuit design sequence.
d. Describe a bipolar integrated circuit processing sequence.
e. Describe the BIMOS process.

## Self-evaluation Questions

*Watch for the answers to these questions as your read the chapter. They will help point out the important ideas presented.*

a. How many diffusions are needed to form a bipolar transistor?
b. How many diffusions are needed to form an MOS transistor.?
c. Explain the difference between metal-gate and silicon-gate technology.
d. How is CMOS different from NMOS?
e. How many masks are needed in a typical bipolar fabrication sequence?
f. What are the advantages and disadvantages of BIMOS?

## 8.1   TECHNOLOGY FAMILIES

The integration of various device structures such as MOS and bipolar transistors, resistors, and capacitors, has been described in Chapter 7. In fabricating integrated circuits using these components there have emerged specific processes that are favoured at any given time, some having been superseded as the capabilities of the technology have improved. In this chapter we first highlight the problems of integrating MOS and bipolar transistors and then proceed to discuss in detail those which presently have a dominant role in the semiconductor industry.

The favoured technologies or technology 'families' have emerged, partly in response to the advances in technology and also in part because of innovative ideas that have arisen. Some have fallen by the wayside because of some particular limitation. A brief list of the more important technologies is shown in Table 8.1. The major subdivision is based on the type of transistor (MOS or bipolar) used as the active element. The properties of the two types of transistor have an important influence on the relative advantages and disadvantages of the technologies and on how they are implemented. It is instructive therefore to review briefly those properties that have a bearing on their fabrication in integrated form.

Table 8.1   Various technology families

| Bipolar | MOS |
| --- | --- |
| Transistor–transistor logic (TTL) | Metal-gate p-channel (PMOS) |
|  | Metal-gate n-channel (NMOS) |
| Emitter-coupled logic (ECL) | Silicon-gate NMOS |
|  | Silicon-gate CMOS |
| Integrated injection logic ($I^2L$) | Double-diffused MOS (DMOS) |

### 8.1.1   The bipolar technology

Figure 8.1(a) shows the bipolar transistor in its simple form. The necessary npn structure in this case is usually formed by two diffusions, one p-type, the other n-type, into an n-type substrate. The major current flow is between emitter and collector and is in a direction normal to the surface (usually termed 'vertical'). One problem is immediately highlighted by this structure. To integrate many such devices on the wafer surface all three transistor connections need to be available on that surface. The collector region in Fig. 8.1(a) is not. This may be facilitated by placing the collector at C' (Fig. 8.1(a)) in which case the current takes the path shown by the dotted arrow. Additional resistance in the collector circuit is however incurred as a penalty. As we shall see in Section 8.3 the solution to this problem leads to greater process complexity.

A second difficulty with the structure of Fig. 8.1(a) is that if more than one transistor were formed on the surface they would not be electrically separate. It is necessary, therefore, in bipolar technologies to provide separate isolation for each transistor as was discussed in section 7.3. This also means greater process complexity and the use of more silicon surface area. A further limitation to the component packing density in bipolar circuits occurs because of the relatively large areas that resistors take up on the surface. The end result is that the circuit packing density in bipolar circuits is not generally as high as for MOS.

However, the bipolar circuit, and in particular ECL, is capable of very high-speed operation and is superior in this respect to MOS circuits. Thus, the general conclusion is that the bipolar circuit is used where high-speed performance is important, and where the number of components per single-chip

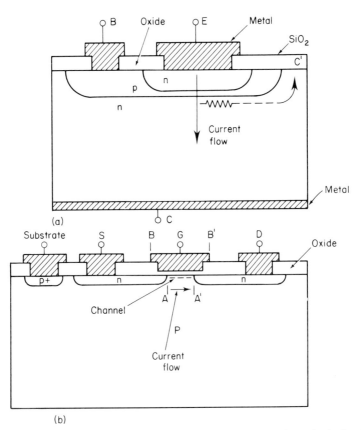

Fig. 8.1  Basic forms of the two transistors used in integrated circuits: (a) bipolar; (b) MOS

circuit is not required to be too high. MOS circuits, as we shall see, are capable of very high packing densities but do not have the high-speed capability of the bipolar approach.

An exception to this general conclusion is the $I^2L$ (Integrated Injection Logic) bipolar technology. Here very high component packing densities can be achieved by (a) eliminating the need for large-area resistors, and (b) by 'merging' transistors so that many occupy the area previously consumed by only one.

## 8.1.2  Metal gate NMOS technology

A simple form of the MOS transistor is shown in Fig. 8.1(b). Depending on the potential applied to the gate (G) an n-type layer is induced in the channel region (hence the name NMOS). This connects together the two n-regions forming the source and drain. The resistance of the channel may be controlled by the magnitude of the gate voltage making the device a voltage-controlled resistor. Alternatively, the channel region may exist in either an OFF state, when the induced n-region is not present, or an ON state where a strong n-region is present. The device then is a switch and may be used as a binary logic element. The current flow, as indicated by the arrow in Fig. 8.1(b), is lateral so that the three transistor terminals lie naturally on the same surface in contrast with the bipolar transistor. The whole of the conducting n-region (source-channel-drain) is surrounded by a p-region. This pn junction forms a barrier to the flow of the electrons into the p-region. Thus the device is electrically isolated from other components and there is no need for separate isolation techniques as in the bipolar device.

## 8.1.3  Silicon-gate NMOS technology

It is essential for the correct functioning of the MOS transistor in Fig. 8.1(b) that the channel region extends the whole way between source and drain otherwise the current path would be incomplete. Because of the alignment tolerance between masking levels it is possible that one of the edges of the gate region (B or B') may not overlap the source or drain (A or A'), unless a sufficient margin of error is allowed. Thus, the overlap of the gate over the source and drain regions is greater than is ideally necessary. This leads to larger gate–source and gate–drain parasitic capacitance, as was discussed on page 131. The device will therefore operate more slowly because these capacitances need to be charged or discharged in order to change the state, and will occupy more silicon area.

This problem led to the development of silicon-gate NMOS. This is a variant of the metal-gate NMOS device just described in which the gate electrode is formed from polycrystalline silicon. The latter can withstand high temperatures and is present during the source and drain diffusion. The sideways movement of the impurities at the gate edges ensures that the gate region always overlaps source and drain. The process and the automatic self aligning property of the Si-gate process is illustrated in Fig. 8.2.

Consider a region on a p-type substrate where a gate oxide has been grown. Polycrystalline silicon is deposited and a gate stripe defined photolithographically (Fig. 8.2(a)). A window stripe is opened in the gate oxide, crossing it at right angles. When the gate oxide is etched the gate stripe acts as a mask to the etch and two windows are thus opened (Fig. 8.2(b)), one for the source and one for the drain.

When an n-type diffusion is then carried out the source and drain regions are automatically registered with the gate region, and because of the slight lateral diffusion at the gate oxide edges, overlap of the gate is ensured. Furthermore, the

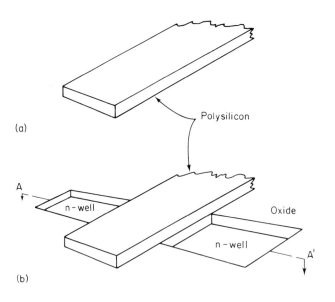

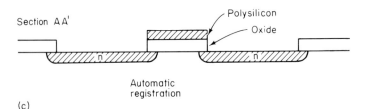

Fig. 8.2  The silicon-gate process: (a) polysilicon gate stripe on oxidized silicon; (b) source and drain window opened; (c) source and drain diffusion

n-type impurities enter the polycrystalline silicon gate stripe, thereby increasing its conductivity. This is course is necessary if the gate stripe is to act as an effective conductor.

The silicon gate, n-channel process is presently the industry standard and in the next section we will therefore describe it in more detail. However, first we need to discuss three other MOS technologies that are of some importance in certain special applications. They are CMOS, DMOS, and VMOS.

### 8.1.4 Complementary MOS (CMOS)

In digital circuits the switching transistors reside in either a 'zero' or a 'one' state corresponding to its being respectively 'on' or 'off'. (The converse may also be true). In the 'on' state the transistor conducts current which gives rise to heat dissipation on the chip. This can be significant when the component density is very high. The CMOS technology, which utilizes a p- and n-channel MOS transistor in the switching element, has the advantage that in either the 'zero' or the 'one' state no significant current flows. Therefore it has very low power dissipation and is used in low-power applications or where heat generation on the chip becomes excessive. The CMOS cell is illustrated in Fig. 8.3. The n-channel transistor has to have p-type semiconductor under the gate and is usually formed by a localized ion-implantation or diffusion.

The drawbacks of this technology are that there are more process steps than for NMOS, and more silicon area is consumed because of the p-well. An additional problem is that there are parasitic bipolars (e.g. npn, see arrow A, Fig. 8.3) and four-layer npnp structures (e.g. arrow B) which can give rise to spurious switching or 'latching' unless designed carefully.

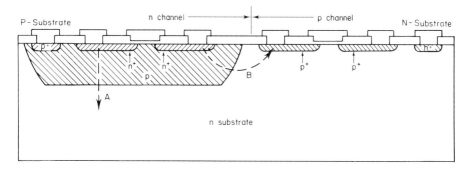

Fig. 8.3 Complementary MOS (CMOS) cell

### 8.1.5  DMOS and VMOS technologies

These technologies are of a specialized nature and have applications in high-voltage MOS devices. Figure 8.4(a) illustrates a DMOS transistor where the channel is defined by a double diffusion. It is possible to achieve very short channel lengths by this technique and to absorb moderate voltages (up to several thousand) across the reverse-biased drain region.

In the VMOS structure shown in Fig. 8.4(b), MOS transistors have been formed on the two faces of a V-groove (see Section 2.4). The channel length is again defined by differential diffusion and high packing densities are in principle possible. A severe problem with this technology is the precise, uniform etching of the V-grooves over large areas of silicon.

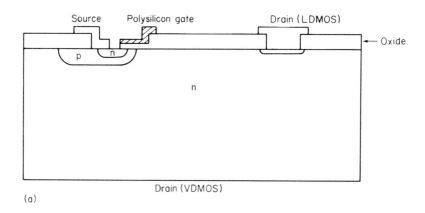

(a)

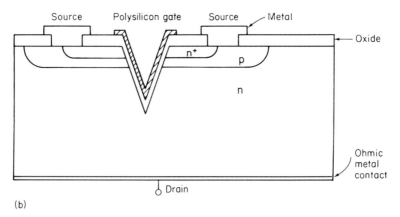

(b)

Fig. 8.4  Basic forms of DMOS and VMOS transistors: (a) double-diffused MOS transistor (DMOS); (b) V-groove MOS transistor (VMOS)

## 8.2   THE SILICON-GATE NMOS PROCESS

In this section we consider the silicon-gate NMOS process in more detail and will discuss the processing of the transistor with its connections, shown in Fig. 8.5. Of course, in a real situation such a transistor would be only a very small part of the whole circuit and the circuit one of many on the mask set. Each stage to be described will therefore actually occur simultaneously over the whole circuit and over the array.

1.  A thick (typically 1 $\mu$m) thermal field oxide is grown in steam over the whole wafer. The first mask (Fig. 8.6(a)) opens a region where the source and drain will be diffused.
2.  It is necessary in some circumstances to adjust the gate voltage at which the MOS transistor turns on. This is done by the gate implant shown in Fig. 8.6(b). The wafer is coated with photoresist and windows opened, aligned as shown with the first mask pattern. The photoresist layer itself acts as a mask

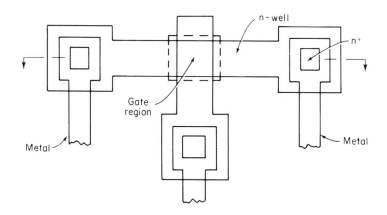

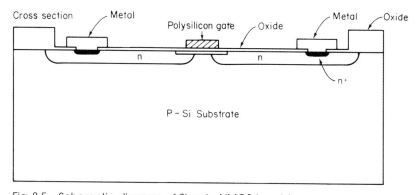

Fig. 8.5   Schematic diagram of Si-gate NMOS transistor

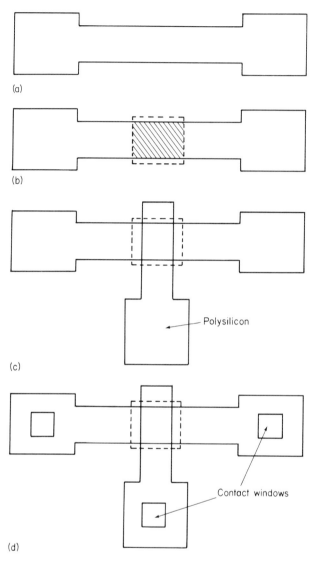

Fig. 8.6  Masking steps for Si-gate NMOS process: (a) diffusion opening; (b) gate implant; (c) polysilicon gate; (d) contact windows

to the implanted ions. Therefore, the only area that receives the implant is that common to both masks (shown shaded).

3.  A thin gate oxide is grown in oxygen over the whole surface and polycrystalline silicon is deposited. The surface is then covered with photoresist and the pattern of Fig. 8.6(c) defined such that the resist is removed from everywhere

except within its boundary. The polysilicon layer is now etched away leaving only those regions protected by the resist.

4.  The wafer is now diffused with an n-type impurity to form the source and drain and to dope up the polysilicon and hence increase its conductivity. The fourth mask is then used to make openings through the oxide for contacts to the source drain and gate regions. These are shown in Fig. 8.6(d).

5.  Finally metal, usually aluminium or aluminium/silicon, is deposited over the surface and the fifth mask used to define the metal pattern to give the final form shown in Fig. 8.5(a). These electrodes provide interconnections to other parts of the circuit.

It is sometimes the case that before the contact openings are made in Step 4, the wafer is coated with oxide. This is to give additional protection to the circuit and to minimize the risk of short-circuits between the metal and underlying regions.

## 8.3   THE BIPOLAR PROCESS

The bipolar process we will describe represents some advance from the standard buried collector process invented in the early 1960s. This technology was satisfactory until the circuit complexity exceeded a thousand components per chip. One improvement over the standard process involved the use of oxide as a means of isolating the transistor laterally, leading to a reduction in the area of silicon used for each transistor and hence an increase in component density.

The bipolar transistor to be integrated is illustrated in Fig. 8.7.

The processing sequence is shown in Fig. 8.8 which corresponds to that for an npn transistor:

1.  A field oxide is grown on a $p^-$ substrate and an opening made using the first mask for the diffusion of a buried $n^+$ collector region. The buried layer has two functions; it isolates the transistor from the underlying $p^-$-region by the $n^+ p^-$ junction formed and gives a highly conducting path for the collector current.

2.  A p-type epitaxial layer is grown and an $Si_3N_4/SiO_2$ composite film deposited using CVD. This film is etched away except over those areas that are to form the base and emitter, and the collector regions. The second mask is used for this step.

3.  Part of the epitaxial layer in the unprotected areas is etched away and $SiO_2$ is regrown. It grows very slowly where the nitride film is (see Chapter 4) and much more quickly in the other areas. Thus it grows downward to join the $n^+$ buried layer completing the isolation of the transistor, and upwards to give an approximately plane surface.

4.  A window is now opened over the collector region to diffuse an $n^+$-contact down to the buried layer (Mask 3).

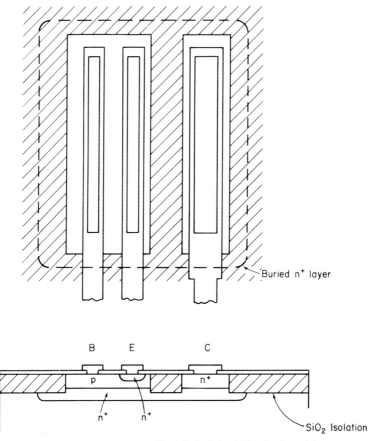

Fig. 8.7   Schematic diagram of oxide-isolated bipolar transistor

5. An opening is made for the implantation of the $n^+$-emitter (Fig. 8.8($\epsilon$)) using the fourth mask, and the emitter region is implanted with phosphorus.
6. Similarly the base contact mask (Number 5) is used to implant $p^+$-ions in the base contact region. The base itself is the original epitaxial p-layer but to make a low-resistance contact it is necessary to increase the p-type density in the vicinity of the contact.
7. Contact windows for base emitter and collector are opened with the sixth mask (see Fig. 8.8(g)).
8. The entire surface is coated with aluminium and the connection pattern defined with the seventh and final mask.

The technology just described is not the only one in use in the semiconductor industry. Other variants and improvements are explored and possibly used by

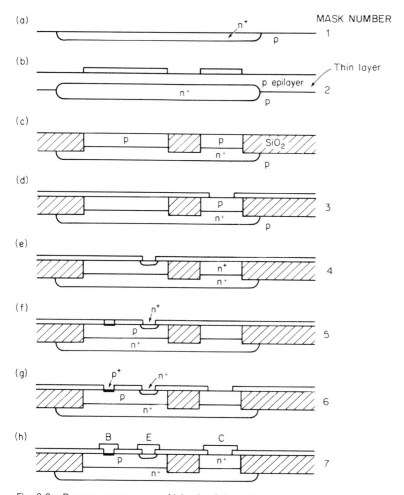

Fig. 8.8   Process sequence of bipolar fabrication

manufacturers in an attempt to obtain the best compromise between performance, complexity, and yield.

## 8.4   INTEGRATION OF A COMMERCIAL CIRCUIT

In this section we illustrate the application of the technologies described in earlier chapters to the integration of a high-performance digital to analogue converter (DAC) manufactured by the Plessey Company.[†]

[†] We are indebted to the Plessey Company, Caswell, United Kingdom, for the use of their DAC design information.

It is not our intention to enter into a detailed description of the function of the circuit as this is beyond the scope of the text. Instead we will discuss the sequence of operations the designer followed in the design of this circuit, then highlight some of the design options he was faced with.

Finally we will illustrate how and where various parts of the circuit actually appear on the chip.

### 8.4.1  The design sequence

The design sequence is illustrated in Fig. 8.9. First the circuit type must be decided. There are several methods of achieving the D/A conversion and the choice has to be made based on a number of considerations. First and foremost, will the technique give the performance required by the specification? Second,

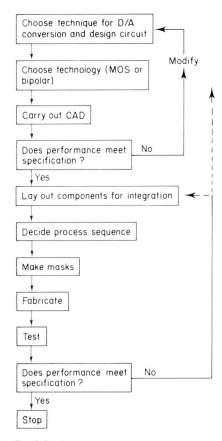

Fig. 8.9   Design sequence for DAC

will the implementation of that technique on chip give rise to any problems? For example, there may be many more components for one technique as compared with the other, or it may contain components difficult to integrate with the required tolerance.

The technique opted for in the case of the Plessey DAC was the so-called 'multiple-current approach'. The circuit is illustrated in Fig. 8.10. It has the property that if each transistor pair is designed identically and placed close together on chip they will be matched so that if similar care is taken over the resistors $R$ they switch equal currents at equal current densities and so will have closely matched speeds.

This would not be possible with a more common technique which uses $R$-$2R$ ladder networks, since each stage operates at half the current of its predecessor. This would give rise to different current densities between stages so that temperature differences arising from internal heat generation would change the operating characteristics of the individual stages.

Following the choice of circuit type a decision has to be made on which technology to use. The first option faced was to use either MOS or bipolar. In the case in question the component packing density is not very high but the performance is crucial: the primary aim in the design is to produce an 8-bit DAC with a settling time less than 5 ns. Such a performance would make it attractive to potential users; if the settling time were 100 ns it would be unlikely to be a worthwhile product. The choice clearly then has to be for bipolar because of its inherently higher speed. A secondary decision to be made is which bipolar technology to use. In this case the demands on speed enforce the use of Emitter Coupled Logic (ECL). In this technology the transistors do not saturate and therefore very low switching times can be achieved. The circuit thus chosen to fulfil the required function is shown in Fig. 8.10.

The next question that arises is whether the circuit as conceived will function with the required specification. Here the designer is faced with three options.

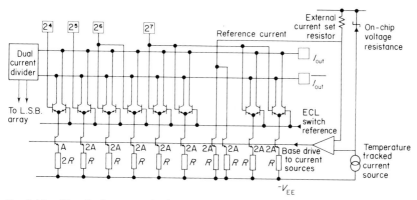

Fig. 8.10   Circuit diagram of D/A converter

Firstly, he could simply proceed through the full intregation sequence and produce chips with the circuit on which he then could test. This is clearly unrealistic because of the great expense in layout, mask-making, and processing (more so if several iterations are required to meet the performance requirements). Secondly, he could build the circuit using discrete components ('breadboarding') and measure its performance. In the present case the operation of the circuit is very sensitive to the layout on the chip so that the designer would be unlikely to get the required information. Thirdly, Computer Aided Design (CAD) could be used. Over the last few years software packages have been developed that will calculate the performance of circuits of moderate complexity very swiftly and economically. This approach has the advantage that various parameters can be adjusted in order to optimize performance. Information may also be obtained on the sensitivity of the performance to small changes in component values that inevitably occur through process variations. In addition, the effects of temperature changes from one device to the next may also be observed. Thus primary performance may be established and a great deal of secondary information may be obtained before the circuit is built.

If initial CAD runs show that the circuit has design errors or does not perform adequately the designer will have to adjust the circuit design, recalculate its performance, and continue this iteration until satisfactory operation is obtained. In the case of the Plessey design, CAD was used to establish that the circuit of Fig. 8.10 operated with sufficient switching speed. The results of the CAD predictions following optimization are shown in Fig. 8.11.

The next question facing the designer is how the components of the circuit should be laid out on the chip. Particular care has to be taken so that the transistors and resistors are closely matched and operate at the same temperature. An indication of the layout is given in Fig. 8.12. The main resistor array has

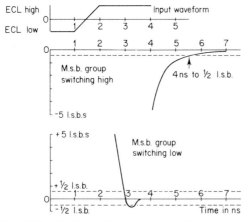

Fig. 8.11    CAD results for current switches

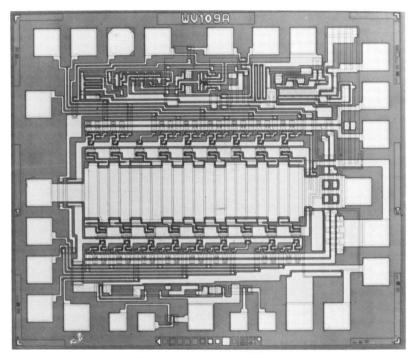

Fig. 8.12  Photograph of DAC chip (courtesy of Plessey Company plc)

been placed at the centre of the chip and the current switches placed above and below them.

Following the layout a detailed process sequence has to be decided. A cross section of a transistor used in the circuit is shown in Fig. 8.13. A thin epitaxial layer is used with ion implants for isolation, the $p^+$-regions, base, and emitter. The minimum feature size is 4 $\mu$m. The design has similarities with the bipolar technology discussed in Section 8.3 but clearly differs in detail. Resistors were implanted in order to achieve the close control over their values that is required. The detailed processing sequence will not be given here as it will be similar to that discussed in Section 8.3.

From a knowledge of the layout and the precise geometry of each masking level, pattern generation tapes are produced containing this data in digital form, which then are used to make the required masks. Finally, the masks are used in the fabrication area and the circuit processed in its entirety. A photograph of the complete chip is shown in Fig. 8.12.

In the case of the DAC of Plessey the measured performance was generally as predicted by the CAD results and fulfilled the performance specification initially aimed for.

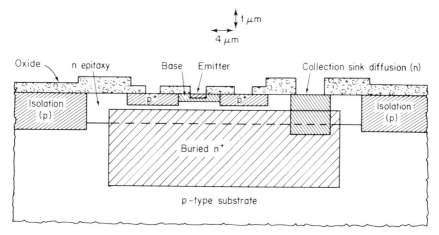

Fig. 8.13   Cross section through bipolar transistor used in the DAC design

## 8.5   USE OF CAD IN VLSI DESIGN

The overall design sequence necessary to produce a finished chip that can be supplied to a customer is now a very time-consuming and difficult one, given the complexity of the circuits involved.

Several tens of man-years of effort might be involved in the design of the more complex circuits and it is therefore highly desirable that the design sequence is augmented by the use of computer tools at all points where it is advantageous to do so.

Many suites of CAD software are now available for designers and in what follow a brief introduction is given to each of the modules in such a suite, indicating how they facilitate the design sequence.

The first and most obvious way in which a computer can help the designer is to enable him to view parts of his design at will, by providing good *graphics*. Working drawings are wholly inadequate as they would contain far too much detail and would be very difficult to change.

Starting with the required function the proposed circuit has to perform, the designer may want to start by breaking it down into simpler sub-systems and then to consider the design of each of these. For a digital circuit he would typically put together a series of logic gates in a certain configuration and would then want to know if this gives him the required function. Here he would use a *logic simulator*. Alternatively, an analogue circuit designer would do the same thing and use a *circuit* simulator to assess the operation of the circuit. At this point several stages of iteration may occur, with the designer changing the initial circuit in order to produce the desired performance.

When the design is to be implemented on silicon, *layout* software may be used to set out the circuit in such a way as to make the best use of silicon area. Interconnections need to be made between components and this can be greatly facilitated by the use of *auto-routing* software which selects appropriate paths for interconnections and indicates where such connections are topologically impossible. The designer would again interact strongly in this part of the design sequence, trying several different layout combinations in order to achieve the most compact. It is clear that the software is used as a tool by the designer enabling him to speed up those parts of the process that are time-consuming or boring, but allows him to exercise his initiative in the design process.

When the layout is satisfactory the designer will wish to set it out as a series of different masking levels appropriate to the particular technology to be used to implement it in silicon. At this stage it is necessary to check whether any of the design rules associated with that technology have been violated. *Design rule checking* software is available for this and may be used by the designer both to indicate the violations and to correct them.

The final way in which CAD may help the design process is by simulating the sequence of processes undergone during fabrication *Process simulation* predicts the effect of diffusion, oxidation, photolithography, deposition and all other fabrication steps on the final form of the silicon. This type of simulation is more appropriate to fabrication engineers than the circuit designer and is presently in a relatively primitive state of development.

## Problems

8.1   Describe the operation of a basic MOS *n*-channel enhancement mode transistor.

8.2   Explain the differences between metal-gate and silicon-gate MOS technology.

8.3   Explain the term 'self-aligned gate'.

8.4   What are the disadvantages of CMOS over MOS?

8.5   How many steps are typically involved in designing a simple D/A converter, as described in Section 8.4?

# CHAPTER 9

# Future Developments in Semiconductor Microtechnology

## Instructional Objectives

*This chapter discusses the present state of the art of microtechnology and its future developments. After reading this chapter you will be able to explain what the future developments in semiconductor technology might be.*

## Self-evaluation Questions

*Watch for the answers to these questions as you read the chapter. They will help point out the important ideas presented.*

a. What advantages to integrated circuits achieve over vacuum tubes?
b. What is the projected cost per gate for an integrated circuit in 1995?
c. What are the current operating frequencies for microwave field effect transistors?
d. What semiconductor materials are currently being used at frequencies in excess of 100 GHz?
e. What are heterojuction devices?

## 9.1 TECHNOLOGY: THE STATE OF THE ART

The past twenty years have witnessed an unprecendented development of microelectronics, from the simple diode and transistor structures which displaced the once established thermionic valves, through simple integrated circuits (ICs) to the staggering complexity and density of 'state of the art' large-scale integrated circuits (LSI). Such circuits are currently reaching levels of 1000 000 transistors per silicon chip. The significance of these rather large figures may be hard to fully appreciate for those students who have no familiarity with thermionic valves and their associated power supplies, and so cannot make a comparison. At this juncture it should perhaps be emphasized that integrated circuits achieve much more than the obvious physical miniaturization of component volume. Four further significant advantages accompany their use; these are:

a drastic reduction in power consumption;
an improvement in reliability;
low cost;
high speed;

these three features cannot be over-emphasized. The reduction in power consumption, for example, although not as easy to visualize, is as large as the physical reduction in circuit size. The transistor has no need for the heater power supplies which are central to the operation of thermionic valves and which represent a 100% power loss. The valves are in fact much worse than this, since in the early valve-operated computers, the power generated for the thermionic process had to be removed to avoid overheating and this of course demanded the operation of powerful cooling fans and resulted in further wasted power.

With regard to reliability, there are two points that need to be emphasized. In the first instance, solid state transistors, operating at around room temperature, have much longer life than that of the heated filament thermionic valves. This factor alone added significantly to computer reliability (and would indeed to any electronic circuit!). However, the current development of the monolithic integrated circuit in preference to solder-interconnected discrete components made further and dramatic improvements to reliability. The soldering of component devices is a relatively primitive technology—very prone to soldering defects (dry joints) and, because it is labour intensive, it is relatively expensive. In contrast, current *planar silicon* technology is a clean, easily automated, and precise technology. This has led to the large reduction in real cost, an increase in circuit yield and trouble-free operation of electronic circuits. This, together with the complexity and versatility of ICs, has resulted in a situation where ICs are now a major influence in our working and leisure lifestyle.

One of the most important applications of silicon ICs—the microprocessor—has now permeated very many aspects of our leisure and work activities and indeed the potential for robotics opened by ICs in industry is at least for the immediate future a very worrying prospect for those people forced out of employment.

To illustrate the important achievements in cost reduction and reliability we show some relevant data in Figs. 9.1 and 9.2. Figure 9.1 shows the cost per gate[†] versus time from 1970 through to 1990. From the logarithmic cost scale we see that

---

[†] A gate is a term for a logic circuit component consisting of roughly four transistors.

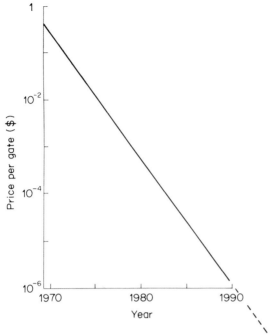

Fig. 9.1   A plot of the cost per gate versus time from 1970 to 1990 (after Roberts, Chapter 1 in Ref. 5)

the reduction from \$1 per gate to (\$10$^{-6}$) per gate (i.e. a reduction factor of 1000 000) in just 20 years is as significant as reduction in the size and power waste.

If we look at the reliability aspects, then some further information emerges. This is illustrated in Fig. 9.2, which shows a schematic curve of the failure rate per function versus the scale of integration. As the scale of integration is

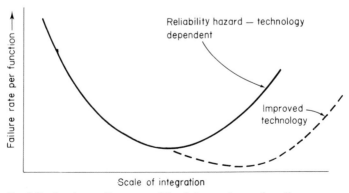

Fig. 9.2   A schematic curve of the failure rate per function versus scale of integration

increased there is an initial fall in the failure rate per function because more of the total circuit is placed on a chip, thus decreasing the number of unreliable interconnections associated with an assembly of discrete devices or small ICs. This improvement does not continue indefinitely since a point is reached where the complexity is so great that the creation of new reliability problems dominates the circuit performance. One example could be the result of the sheer size of the chip. The larger the chip area, the greater the chance of its containing a destructive structural defect, as was suggested in Chapter 1 (Fig. 1.1). Consequently, this demands improvements in the growth and preparation of the starting wafers. Another example of this problem is the attempt to increase the scale of integration by reducing the component sizes. At any particular time there will be a limit to technological processes such as the resolution of the lithography. Attempting to push these to their limits will result in a degradation in component quality and hence a reliability–yield problem.

These few examples serve to emphasize the point that the optimum in scale of integration shown in Fig. 9.2 is not permanent limitation but is strongly a function of the technology in question, or at least of its weaker points.

## 9.2  FUTURE TECHNOLOGY DEVELOPMENTS

To the semiconductor industry wishing to ensure a future for itself in the 1990s the assessment of future developments and trends in semiconductor and IC technology is of prime importance. The rapid and sometimes spectacular developments made during the last two decades make this type of prediction very difficult indeed. Having made that point, it must however be emphasized that much of the progress is the direct result of wise investment in research and development, both in terms of manpower and equipment. This trend will continue but it must be remembered that current and future developments are very expensive and to a great extent demand investment decisions outside the domain of the scientist and a engineer and in the hands of Government and large industrial corporations.

What then are these technology developments likely to be? Some clearly identified regions are as follows:

(a)  higher packing density and greater complexity and VLSI;
(b)  higher operating speeds;
(c)  new semiconductor materials and devices.

Fairly clearly, the division presented above is somewhat simplified, since many of the objectives overlap. The first of the three points is at the present time the most actively being studied. It is centred on silicon, the 'bread and butter' material of the industry. Current development directed at (a) will achieve some of the objectives of (b). The technology may translate, with modifications, to any new semiconductors developed in the future (i.e. (c)).

### 9.2.1    Ultra large scale integration

An important current development is the drive to Ultra Large Scale Integration (USLI). Central to this development is the need to decrease component size, with the objective of working with device dimensions in the *submicron* regime. The way in which device dimensions have decreased in size, and how they are likely to in the future, is reflected in Fig. 9.3(a). This 'law' of growth was first mentioned by Gordon Moore in the early days of IC development, since when it has become known as Moore's Law. This size reduction, and its dramatic effect on cost and performances, has resulted in the remarkable growth in the production of ICs, as indicated in Fig. 9.3(b), which compares the market size in IC production between the USA, Japan, Europe, and the rest of the world over the period 1981 to 1987. The break point at around one micron is significant since it represents the approximate limit of conventional optical lithography and therefore is the limit of current technology. The two approaches which overcome this limit are *electron beam lighography* and *X-ray lithography*. In both cases the resolution limit caused by the wavelength of ultraviolet light can be overcome. No industrial semiconductor manufacturer wishing to stay in business can affort to ignore these developments in spite of the fearful cost of research.

The submicron developments will achieve two main objectives: A technology for ULSI and higher operating speeds for conventional LSI circuits. The higher speeds resulting from the submicron dimensions are also very important for the development of microwave field effect transistors (MESFETs). In this case, the small gate width decreases the electron transit time and enables the transistor to operate at a higher frequency. Submicron dimensions, coupled with the use of the high mobility material, gallium arsenide, are currently producing MESFET devices that can operate at up to around 40 GHz.

As the submicron regime is entered a number of important difficulties arise which have to be considered. First there is the problem of scaling. It is not sufficient just to reduce the surface dimension and leave those normal to the surface unaltered. Nor can smaller devices be operated at the same voltage. Simple constant field scaling predicts that voltages will drop in proportion with gate-length reduction. However, current densities then increase (as the square of the scaling factor). This leads to a limitation set by electromigration in the conductors. Also if operating voltages are reduced, then noise margins are more difficult to maintain.

The second problem that arises concerns the changes in the physics that occur when devices are made sufficiently small. Impurity atoms are sufficiently closely spaced in present-day devices (e.g. for a doping density of $10^{21}$ m$^{-3}$ the mean spacing between dopants is 0.1 $\mu$m) so that when ionized they can be treated as a uniform distribution of charge. However, when device dimensions get to below one micron the granularity of the charges becomes important. Present-day device physics relies heavily on the assumption that electrons and holes are

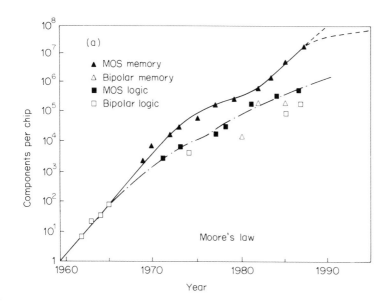

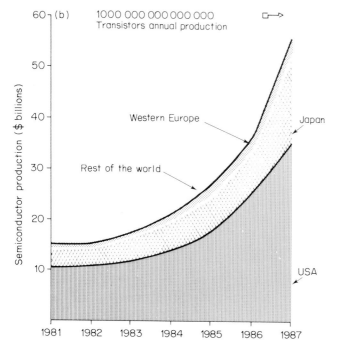

Fig. 9.3  (a) Increase in the number of components per silicon chip since 1959; (b) world-wide production of integrated circuits

scattered many times during single pass through a device. Such scattering may occur from other electrons or holes, impurity atoms, crystal imperfections or phonons. Scattering processes are characterized by the average time between individual such events. When device dimensions become sufficiently small it may be that only a few collisions will occur during the passage of a carrier through a device. In this circumstance the basic transport mechanism is no longer scattering dominated but approaches a ballistic-like motion in which carriers accelerate uniformly in the applied field.

Current mathematical descriptions or 'models' of semiconductor devices are usually one-dimensional. They usually describe their operation with reasonable accuracy while analytical solutions to two- and three-dimensional formulations do not exist except possibly in a very restricted number of situations where the geometries are artificially simple. However, as device dimensions become submicron the simple models are no longer accurate. This means that two-dimensional models have to be used and a resort made to numerical solutions. The latter usually mean very large computation times and a loss of physical insight into the operation of the device.

### 9.2.2  New semiconductors

With regards to new semiconductor materials, this is still a very open question. It is going to be a hard task for any material to displace silicon in terms of cost, availability of base material, and its commanding lead in a highly developed planar technology. At the present time certain III–V semiconductors are being developed for specialized applications where they can improve on silicon. The most important of these is gallium arsenide, followed by indium phosphide. Gallium arsenide was initially developed because it exhibits the transferred electron effect and therefore is used to fabricate two terminal microwave devices (Transferred Electron Devices, or just TEDs). GaAs and InP TEDs are still important and provide the basis of microwave power sources up to and in excess of 100 GHz.

Gallium arsenide MESFETs with their high operating speed are now being used as the basis of a new family of integrated circuits and two specific areas of work are identified:

(a) monolithic analogue microwave ICs;
(b) digital circuits for gigabit operation.

The first of these applications is now well established with a clear role in future microwave systems. The second area of development is still an open question, because although working ICs have been fabricated they still have not achieved the predicted speed improvement over silicon. At the present time the major investment in this work is military in origin.

The III–V semiconductors and their ternary and quaternary compounds have a clear role in producing optoelectronic devices, lasers, light-emitting diodes

(LEDs), photodetectors, and integrated optical systems (the optical sequel to the conventional IC). The rapid development and use of optical communication systems has established a clear role and future for these materials. The use of long wavelength optical devices (i.e. $\lambda \simeq 1.3 \, \mu$m) has enabled optical communication systems to operate without repeater stations in links up to 100 km in length. The move from concept to real systems has taken place in just over 10 years, a remarkable achievement!

### 9.2.3  Heterojunction devices

The development in recent years of sophisticated epitaxial crystal growth technique such as MBE and MOCVD (Sections 2.2.8 and 2.2.9) has led to the fabrication of very advanced devices which utilize the properties of heterojunctions. The term 'homojunction' is used for a system such as a pn junction in the same material as shown in Fig. 7.1. In contrast a heterojunction involves two different semiconductor materials, as for example in a silicon to germanium transition. Another important heterojunction is that between GaAs and AlGaAs and this provides the basis of two important engineering examples that will be described in this section. It should be stressed that there are now very many new and innovative devices being researched in laboratories throughout the world. Here we will discuss only three examples which build upon the discussion in the early part of this book. The examples concerned are the High Electron Mobility Transistor (HEMT for short), the Heterojunction Bipolar Transistor (HBT) and the double heterojunction laser.

*High electron mobility transistor*

Figure 9.4(a) shows an n-type doped AlGaAs surface in contact with an undoped GaAs surface. After equilibrium the band structure relaxes to the form shown. At the interface the small potential well forms in the GaAs conduction band and this fills with electrons from the doped AlGaAs. Now the interesting feature of this conduction band is that electrons move into this well, and, provided the small undoped spacer layer ($\simeq 40$ Å) is grown at the AlGaAs side of the interface, the electrons do not have to contend with scattering from positively charged donor atoms which would reduce their mobility. The important consequence of this is that if the electrons in this 'two-dimensional well' (a low-dimensional structure—hence the term LDS) are subjected to an electric field along the layer their mobility corresponds to that in undoped GaAs and is much higher than in a layer doped to $10^{23}$ m$^{-3}$ corresponding to that in a MESFET (see Fig. 7.10).

This high electron mobility layer can be used to fabricate a field effect transistor as is shown in Fig. 9.4. This bears a very close resemblance to the MESFET in Fig. 7.10 and transistors using such a heterojunction have a much

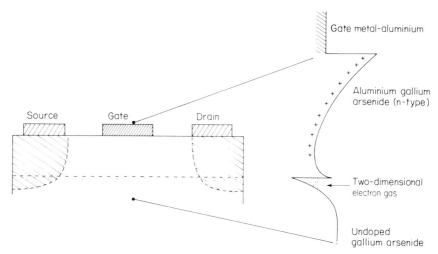

Fig. 9.4    Illustrating how the heterojunction shown in Fig. 2.10(a) is used to fabricate a high-electron-mobility FET

improved high-frequency performance compared to a conventional MESFET. Scientists have already fabricated devices with cut-off frequencies ($f_T$) of over 100 GHz. Both single devices and IC are being developed from this HEMT technology.

### The heterojunction bipolar transistor

A second class of device that has also already shown significant engineering promise is the HBT. The device is capable of being designed to have an exceptionally high current gain or, alteratively, gain may be sacrificed in order to increase the device $f_T$. At the present time an $f_T$ of over 100 GHz has been achieved. A conventional 'vertical' planar transistor is shown in Fig. 7.8. (It must be remembered that the vertical scale is greatly magnified for clarity.) In order to achieve useful current gain the emitter is heavily doped ($\sim 10^{25}$ m$^{-3}$) and the base lightly doped ($\sim 10^{23}$ m$^{-3}$). That is necessary because the current gain ($\beta$) is

$$\beta \propto \left\{ \frac{N_D(\text{emitter})}{N_A(\text{base})} \right\}. \tag{9.1}$$

A second condition that is needed is for the base region to be very thin (i.e. compared to the electron diffusion length in the base). A consequence of these two design criteria is that the base parasitic resistance increases when the transistor is designed for increased current gain. Gain is increased at the expense of high-frequency performance—this is true because the emitter base junction has to be charged and discharged through the base resistance and hence an increasing base resistance slows up the device frequency response. This difficulty is overcome in the HBT in the following way: a junction between an n-type doped AlGaAs and a p-type GaAs is shown in Fig. 2.10(b). If the junction is graded so that the Al concentration is increased gradually from the GaAs to the final AlGaAs alloy ($\sim 40\%$Al, $60\%$Ga) then the conduction band notch does not occur (Fig. 2.10(c). When this is now fabricated into a transistor structure as in Fig. 9.5, the potential hill facing the electrons going from emitter to base is smaller than that experienced by the holes going from base to emitter. This example of 'band gap engineering' selectively inhibits the unwanted base to emitter hole current and increases the emitter efficiency and the current gain $\beta$. It can be shown that there is now an enhancement factor in the current gain, i.e.

$$\beta \propto \exp\left(\frac{\Delta E_{g}}{kT}\right) \tag{9.2}$$

where $\Delta E_{g}$ is the differences in band gap between the AlGaAs and the GaAs (note $\Delta E_{g}$ is controlled by the concentration of Al in the AlGaAs layer). Thus for a value of $\Delta E_{g} \simeq 0.25$ eV and at mean temperature $kT \simeq 0.025$ eV, the factor in equation (9.2) is $\simeq 10^{5}$.

Using this principle transistors have been fabricated with a current gain in excess of 5000. Since in most applications this very high gain is not needed the design can be modified to trade gain for improved $f_{T}$. One way of doing this is to increase the value of $N_{A}$ (base) $\gtrsim N_{D}$ (emitter). A base doping reduces the parasitic resistance and consequently increases $\chi_{T}$. Using this trade-off transistors have been fabricated with $f_{T}$ of greater than 100 GHz. In silicon the most promising implementation of the HBT is the Si–Si$_{1-x}$Ge$_{x}$–Si structure. Current gains of 5000 and values of $f_{T}$ in excess of 30 GHz have been achieved at the time of writing.

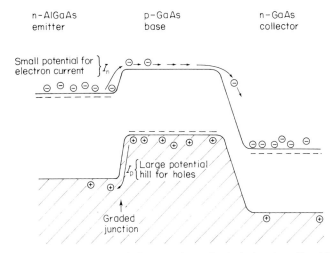

Fig. 9.5   Energy band diagram for a single heterojunction bipolar transistor (see Fig. 2.10 (c))

### The double heterojunction laser

A pn junction when forward biased may emit light and with the appropriate feedback can be converted into a laser. A certain minimum current, the threshold current, is required for laser action to start. For a conventional pn junction laser this threshold current is quite high ($\sim 10^9$ A/m$^2$). Under these conditions the device is electrically stressed, leading to device degradation which limits the lifetime. More efficient laser action can be achieved by using the double heterojunction shown in Fig. 9.6.

The laser action in the double heterostructure device is due to the recombination of electrons and holes in the thin GaAs layer, and typical threshold currents below $10^7$ A/m$^2$ have been achieved. The electrons and holes in this thin GaAs layer cannot move into the AlGaAs because of the potential barriers at the interfaces. The abundance of electrons and holes in these wells allows laser action to take place at much lower diode current densities.

There is a remarkable property of electrons and holes confined to potential wells as is shown in the simplified Fig. 9.7. When the well thickness is reduced to around 10–100 nm the quantized electron energies take on the discrete nature illustrated. The narrower the well dimension $d$, the larger are the minimum energies needed to break a GaAs bond. The minimum energy of an electron in the conduction band is given by the formula

$$E = \left( \frac{nh^2}{8md^2} \right),$$

and a similar formula for the holes in the valence band, where $h$ is Planck's constant, $m$ is the effective mass of the electron (or hole) and $n$ is an integer = 1, 2, ... etc. It will be seen that the energy release when an electron–hole pair recombine is larger than the bandgap $E_g$ and can be controlled by the well spacing $d$. Furthermore, unlike the conventional pn junction where photons of a range of frequencies are emitted, here the photons emitted are uniquely defined by the difference between quantized energies and hence give a sharper line in the

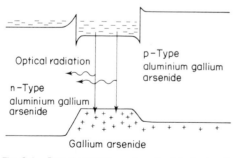

Fig. 9.6   Band diagram of a double heterojunction laser

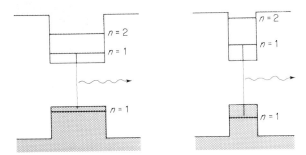

Fig. 9.7  Illustrating how the thickness of a double heterojunction potential well affects the energy difference of the electron energy levels which in turn controls and varies the energy (i.e. frequency) of the light emitted on recombination

laser output frequency. This is put to good advantage in modern optical communication systems.

The three examples described here are important becuase they have already proved their value to microelectronics—they are not devices of the future but ones that are already having a significant impact on our engineering applications, particularly high-frequency engineering.

### 9.2.4  Postscript

An appraisal of the development thus far indicates that progress has been achieved by continuous technological progress punctuated by discontinuous steps resulting from the discovery of a completely new device. Two very good examples of the latter are the discovery of the transferred electron device mentioned earlier and the charge coupled device (CCD), discussed in Chapter 7. CCDs are very interesting because they link the two types of progress. The initial concept of using the transfer of a packet of charge rather than current as the basic unit measure was an original innovation. However, the rapid development and system application of these devices in silicon is a direct result of being able to use the existing planar technology. Herein we have a very *important lesson: When we go away from silicon, the development of engineering quality devices is retarded because of the expense and time needed to establish a range of new technologies.* Hence gallium arsenide and the example of the CCDs made from this material have a long way to progress (if they ever do) to match that of present-day silicon CCDs.

Another important problem concerns the design of integrated ciruits when the component density is very large. It is estimated that, subject to previous experience and earlier designs, several man-years of effort is required to design circuits that are relatively complex but by no means unachievable by present-day standards. The projected design costs of circuits that may perhaps be several orders of magnitude more complex become very large and could become prohibitive. Further, there is the problem of testing such circuits once they are made. There is no way that this could be done manually, even at present, and in the future we are likely to see automated test equipment evolve in sophistication to cope with this difficulty. Such costs must of course be added to the final retail price of the circuit and may have a profound effect on overall economic decisions.

Finally there is the problem of need. There must come a point in the advancement in technology where the capability of the circuits match the requirements of the application. A specific example concerns the home computer market. More and more computing power is presently being given to the home-computer owner. When systems exist that will fulfil everything he is likely to require, he will see little point in purchasing yet more computing power—he won't know what to do with it!

There is of course no way in which we can predict when this will occur, nor is it our intention even to attempt it. However, an eventual saturation in demand is inevitable in the long term and this will have perhaps the most profound effect of all on the semiconductor industry which for the last twenty years, and for an unforeseeable number in the future, has been firmly based on a policy of very rapid expansion.

The story of semiconductor technology is not all rosy—many blind alleys are entered. In terms of devices a good example is the tunnel diode. This is a two-terminal device which can operate at microwave frequencies. However, the combination of difficulties in making devices with reproducible characteristics and the low output power has led to the cessation of the production of these devices. It must however be remembered that these devices added much to our physical understanding of semiconductors and this can be thought of as an investment in the future rather than wasted effort.

The prediction of future developments in semiconductors is no easier today than it was fifteen years ago and very few people then were able to predict the exciting development we have witnessed since that time.

## Problems

9.1  What is the cost reduction factor per gate that the industry has seen over the past 20 years?

9.2  What are the technology development areas likely to be in the future?

9.3  What is the device dimension region for ULSI?

9.4  What is the current operating frequency limit for MESFETs?

9.5  What is the current operating frequency limit for TEDs?

9.6  What is the advantage of a HEMT over a MESFET?

9.7  Why is the gain of an HBT intrinsically higher than that of a homojunction bipolar transistor?

9.8  How is the gain of an HBT traded for higher-frequency performance?

# APPENDIX **1**

# Material and Device Evaluation Techniques

## A1.1  INTRODUCTION

Semiconductor material prepared for a specific device application requires careful evaluation to see if its parameters lie within the specification for a particular device structure. There are available today a wide range of material and device evaluation techniques, with which to characterize both the base material and in some cases to evaluate the actual layers within device structures. In this brief summary of the major techniques, attention will be confined only to the basic principles and no attempt will be made to describe the details of practical systems, since in some cases these are very complex (for example SIMS) and most people are usually concerned only with their application.

For device grade semiconductor layers, the following information is usually required:

(a)  inspection of layer quality, morphology, uniformity, and thickness;
(b)  evaluation of layer resistivity (or conductivity) and mobility;
(c)  measurement of doping profiles.

Ideally it is desirable to conduct these measurements on the actual layers being used for the device fabrication. This is not always possible and sometimes test samples have to be grown alongside the device layer.

### A1.1.1  Layer thickness: angle lap and stain procedure

This technique is suitable for estimating the depth of a pn junction. The basic idea is illustrated in Fig. A1.1 where the junction depth $d$ is small. The wafer is

173

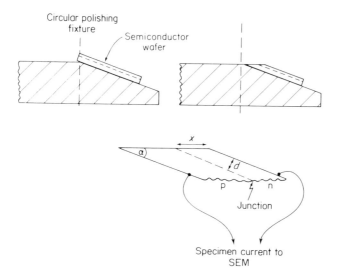

Fig. A1.1   The measurement of junction depth using the angle lap technique. In this example the distance $x$ is measured using an SEM in the specimen current mode

angle lapped by mounting the wafer on a special fixture which allows the edge of the wafer to be wedge polished at a very small angle (1–5°). The small angle taper $\alpha$ enables the distance $x$ to be magnified with respect to $d$; that is,

$$x = d/(\sin \alpha) \gg d.$$

Knowing $x$ and $\alpha$, $d$ can be calculated.

In order to be able to see the pn junction, it is delineated by the use of a special etch that will selectively stain either the p- or the n-region. A useful example for silicon is the chemical etch

(concentrated hydrofluoric acid + (0.1 to 0.5) % nitric acid)

which acts to darken (stain) p material.

*Note:*   Hydrofluoric acid is a hazardous chemical to use and care must be taken in its use!!

An alternative procedure is to use a scanning electron microscope (SEM) to delineate the junction. This is shown in Fig. A1.1, where terminal current across the pn junction is only produced when the actual scanned beam is hitting the junction depletion region. In these circumstances electron-hole pairs generated in the depletion are swept out by the internal field and produce the terminal current. The SEM image is produced from the specimen current (i.e. the so-called specimen current mode of operation).

### A1.1.2   Measurement of layers sheet resistance

The four-point probe technique can be used to evaluate the sheet resistivity of a layer of semiconductor of known thickness which in the analysis is assumed to be homogeneous in its electrical properties. Although it can be used on a homogeneous self-supporting layer, in practice, it is usual to be concerned with a thin near-surface layer electrically isolated from the bulk sample. In the case of a thin p-layer on an n-type base (or the alternative configuration, an n-layer on a p-type base) the electrical isolation is provided by the pn junction. In the case of gallium arsenide the p- or n-layer is usually grown (or implanted) onto a semi-insulating substrate.

### A1.1.3   Specification of sheet resistance

Consider the thin sample of material specified in Fig. A1.2. If the resistance is measured between surfaces A–B then the resistance can be written as:

$$R = \frac{\rho(s)l}{ws} \qquad (A1.1)$$

where the specific resistance $\rho(s)$ is assumed to be uniform across the layer and measured in (ohm-m).
   Equation (A1.1) may be written:

$$R = \frac{\rho(s)}{s}\left(\frac{l}{w}\right) = \rho_\square\left(\frac{l}{w}\right), \qquad (A1.2)$$

where $\rho_\square$ is defined as the sheet resistance

$$\text{Its units are } \frac{\text{ohm-m}}{\text{m}} = \text{ohms}$$

   Note that for a sample with unit sides $l = 1$, $w = 1$, $R = \rho_\square$ regardless of dimensions and hence the use of the term 'ohms per square'.

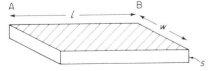

A
B
$l$
w
s

Fig. A1.2

## A1.1.4  Four-point probe

The probe is a practical and simple way of determining the sheet resistance. Here four equally spaced collinear probes are placed onto the layer surface (Fig. A1.3(a)). A current of $I$ (amperes) is passed through the outer probes and a voltage $V$ (volts) is measured using the two inner probes. Note that these two inner probes do not draw a current and there is, therefore, no need to worry about any contact resistances. A value of current $I$ is chosen such that $V$ and $I$ are linearly interrelated.

For situations where the layer is:

     p on n    or    n on p    or    p (or n) on semi-insulating layer

all the current drawn passes in the layer being measured. Now for the outer two probes drawing a current $I$ (Fig. A1.3(a)) the potential at some arbitrary point $P$ is:

$$V_p = \frac{I}{2\pi} \rho_\square \ln\left(\frac{r_2}{r_1}\right) + A,\tag{A1.3}$$

where $A$ is a constant of integration.

Apply this equation to the central electrodes of the four-point probe:

$$V_1 = \frac{I\rho_\square}{2\pi} \ln 2 + A\tag{A1.4}$$

$$V_2 = -\frac{I\rho_\square}{2\pi} \ln 2 + A,$$

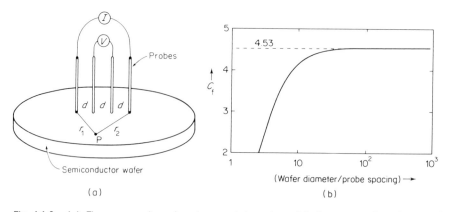

Fig. A1.3  (a) The geometry of a four point probe; (b) the correction factor $C_f$ (equation (A1.6)) plotted as a function of the ratio of wafer diameter to probe spacing to obtain the potential $V$ between the inner probes

$$V = V_1 - V_2 = \frac{I\rho_\square}{\pi} \ln (2).$$

Rearranging,

$$\rho_\square = \left(\frac{\pi}{\ln (2)}\right)\frac{V}{I} = 4.5324(V/I). \tag{A1.5}$$

The equation for $\rho_\square$ presented above assumes an infinite wafer. In practice, the use of finite wafers requires a correction factor $C_f$ where $C_f$ is defined by

$$\rho_\square = C_f(V/I). \tag{A1.6}$$

For the simple case of circular wafers with central probes, the relationship between the correction factor $C_f$ and the wafer diameter/probe spacing is shown in Fig. A1.3(b).

### A1.1.5   Mobility and the Hall effect

The basic principle of the Hall mobility measurement is illustrated in Fig. A1.4, which shows the situation for an n-type semiconductor. The Lorentz force on the electrons deflects an excess of electrons onto the lower surface, and a deficiency of electrons (positive charge) on the top surface. In equilibrium an internal electric field $F_y$ is set up to balance the Lorentz force and hence the net force in the $y$ direction to zero. Generally

$$\mathbf{f} = -q(\mathbf{\bar{F}} + \mathbf{\bar{v}} \wedge \mathbf{\bar{B}})$$

since

$$F_y = v_x B_z.$$

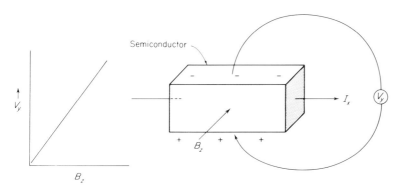

Fig. A1.4   A simple geometrical arrangement for the Hall experiment together with the Hall voltage $V_y$ plotted as a function of the magnetic field $B_z$

Furthermore since the current density $J_x = -nqv_x$

$$\frac{F_y}{J_x B_z} = -\frac{1}{qn} = R_H \qquad \text{(Hall coefficient)}. \qquad \text{(A1.7)}$$

The sign of the Hall voltage determines the nature (electrons or holes) of the charge carriers and their density $n$. Since

$$\frac{J_x}{F_x} = -nq\left(\frac{v_x}{F_x}\right) = -nq\mu_H$$

the mobility $\mu_H$ can be determined.

From the point of view of practical measurements, the Van der Pauw clover leaf sample proves very convenient to measure the mobility and resistivity of a given layer. The active layer is cut into the clover leaf shown in Fig. A1.5. Note that in the case of test samples grown on a semi-insulating (or a blocking pn contact) substrate, the clover leaf shape is only cut into the epitaxial layer. With the appropriate masks, this can be done either by 'mesa-etching' or by sandblasting. The four contacts at A, B, C, and D are ohmic contacts. If it is assumed that the clover leaf sample is uniform and has only one carrier type, then the Van der Pauw analysis yields[†]

$$\rho = \frac{d\pi}{2 \ln (2)} \left(\frac{V_{DC}}{I_{AB}} + \frac{V_{BC}}{I_{AD}}\right) \qquad \text{(A1.8)}$$

and the Hall mobility

$$\mu_H = \frac{d}{B_0 \rho} \left(\frac{V_{CA1}}{I_{BD1}} - \frac{V_{CA0}}{I_{BD0}}\right), \qquad \text{(A1.9)}$$

where $B_0$ is the magnetic field perpendicular to the sample and the extra subscripts 1 and 0 refer to the cases with or without the magnetic field.

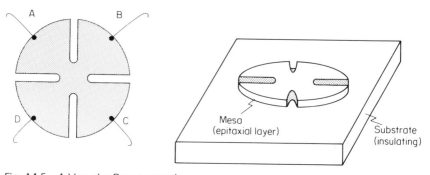

Fig. A1.5    A Van der Pauw sample

[†]L. J. Van der Pauw. *Philips Research Reports* **13**, 1–9 (1958).

### A1.1.6   Geometric magnetoresistance

In the Hall effect the magnetic field was used to set up an electric field which was then used to determine the carrier type, and mobility. An alternative procedure— geometric magnetoresistance—can also be used to determine a magnetoresistance mobility $\mu_m$ which is very closely related in value to the Hall mobility $\mu_H$. For this measurement the active layer geometry must have a large aspect ratio. In its simplest form a circular sample diameter $D$ and thickness $d$ is placed between two ohmic contacts. Here the aspect ratio $(D/d)$ must be $\gg 1$. If this layer is now placed inside a magnetic field $B$ orientated as shown in Fig. A1.6 and the low electric field current density $J$ determined as a function of $B$, then the simple relationship[‡]

$$\frac{J(B = 0)}{J(B \neq 0)} = 1 + k_s^2 \mu_m^2 B^2, \tag{A1.10}$$

where $k_s^2 \simeq 1$. Hence $\mu_m$ can be determined.

In its simplest form, the large aspect ratio inhibits the build up of the Hall field and the charge carriers are forced to take a longer path (a–c, Fig. A1.6) from one electrode to the other and hence the resistance of the sample increases with increasing $B$.

One way of using this technique is to grow a special layer alongside the layer whose mobility is required and to use a heavily doped $n^+$-substrate—this helps to ensure a good ohmic contact to the back. Another important feature of this technique is that it can be used to measure the actual mobility of certain device

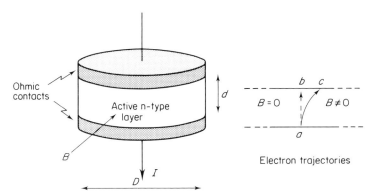

Fig. A1.6   A simple device structure for geometric magnetoresistance (GMR) measurments (i.e. $D/d \gg 1$)

[‡]For a full discussion of this theory see T. R. Javis and E. F. Johnson, *Solid State Electronics*, **13**, 181 (1970).

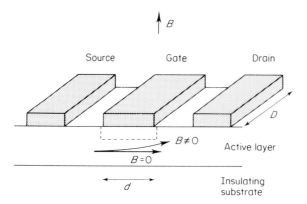

Fig. A1.7    The GMR mobility measurement on an FET device

structure. There are transferred electron devices (Fig. 7.23) which have the simple geometry required by Fig. A1.6. Typically $D \simeq 200 \ \mu m$ and $d \simeq 10 \ \mu m$. Similarly the field effect transistor can also be used to determine $\mu_m$ as is shown for an n-type MESFET in Fig. A1.7.

### A1.1.7    The measurement of doping profiles

One of the most important features to emerge from any discussion of semiconductor technology is the importance of the control and measurement of impurity profiles. They define the device type and control its performance. In very simplistic terms there are two general types of profiles that may be measured: (a) the electrically active profile and (b) the total impurity profile. In the former case, only those impurity atoms that are electrically active contribute (for example correctly located p- or n-type dopants and deep level trapping atoms), whilst in the latter case the impurity atoms may be electrically inactive. From the point of view of device performance the first category is the important one but information on the second type of profile is frequently needed in order to determine what percentage of atoms have become electrically active. Some of the most important techniques are summarized in Table A1.1.

Table A1.1

| Technique | Type of information |
|---|---|
| Radioactive isotopes plus stripping | Total (b) |
| SIMS | Total (b) |
| Four-point probe plus thinning | Electrically active (a) |
| Hall measurements plus thinning (Clover leaf) | Electrically active (a) |
| Capacitance-voltage technique | Electrically active (a) |

In the category (b), the techniques provide a measure of the total concentration of dopant (or impurity) in question and do not differentiate between electrically active atoms and others.

## A1.1.8   Radioactive isotopes

This technique is limited to dopants which have a radioactive isotope with a half-life in a suitable range. The basic principle is as follows: radioactive isotopes are diffused or implanted into the semiconductor. If the total number of atoms is $Q$ then the use of a suitable nuclear particle detector will enable the degree of radioactivity $R$ to be measured. For a given measurement time interval $R$ will be a function $R(Q)$ of $Q$. If a thin surface layer, thickness $\Delta x$, is removed and the new activity $R$ measured, $R(Q - \Delta Q)$, where $\Delta Q$ is the number of isotopes in the removed layer $\Delta x$ and is the profile information. That is, using $R_0, R_1, \ldots, R_n \ldots$ to refer to the sequence of measurements.

$$Q_1 \propto R_0 - R_1$$
$$Q_2 \propto R_1 - R_2$$
$$\vdots$$
$$Q_n \propto R_{n-1} - R_n$$
$$\vdots$$

Using this technique, the differential profile can be built up. Calibration of the profile can be achieved by implanting a known quantity of ions into the solid and measuring their activity. With regard to the constraint on half life, this must be short enough to provide a measurable signal in a finite time for doping levels in the range $10^{21} - 10^{24}$ m$^{-3}$ but not so short that the radioactive decay of atoms during the measurement is large compared to the initial number implanted.

## A1.1.9   Secondary Ion Mass Spectrometry (SIMS)

This is a specialist technique that requires expensive equipment. In its simplest form, it consists of an ion beam projected onto the solid surface (Fig. A1.8). The use of relatively lower energy beams of 5–40 keV, oxygen or cesium ions causes the surface to be eroded away slowly (this process is called sputtering). If the sputtering is controlled to occur at a given constant rate, and the concentration per unit time of the specified sputtered ions is measured by mass spectrometry analysis, then the required profile can be obtained. The sputtering time scale is converted into a depth scale by measuring the total crater depth in a given time interval. This technique provides a relative profile which can be calibrated by using known implanted test samples.

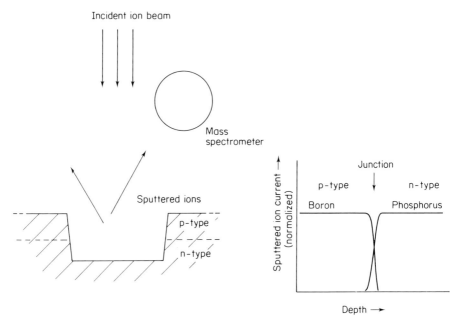

Fig. A1.8    Illustrating the principles of the secondary ion mass spectrometry

### A1.1.10    Four-point probe

The four-point probe technique coupled with an anodic stripping process can be used to determine depth profiles. Let

$$\rho_1 = \text{sheet resistance before the removal of a layer}$$
$$\rho_2 = \text{sheet resistance after the removal of a layer}$$
$$\Delta x = \text{layer thickness removed.}$$

The average specific resistance $\bar{\rho}$ of the layer removed is given by

$$\bar{\rho} = \frac{\rho_1 \rho_2}{(\rho_1 - \rho_2)} \Delta x \text{ ohms m.} \tag{A1.11}$$

Successive removal of thin layers $\Delta x$ provides $\bar{\rho}(x)$ and hence the doping concentration $N(x)$.

### A1.1.11    Capacitance voltage profiling

For this technique, the layer is grown on a heavily doped substrate and a Schottky barrier placed on the upper surface of the layer, with an ohmic contact

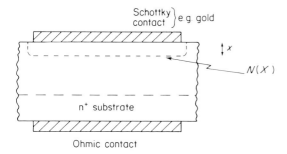

Fig. A1.9   The use of a Schottky diode for capacitance voltage profiling

to the substrate (Fig. A1.9). The basis of the technique is the change in terminal capacitance with applied voltage given by the equation

$$N(x) = -\frac{2}{q\varepsilon_s S}\left[\frac{1}{d(1/C^2)/dV}\right].\qquad (A1.12)$$

where $S$ is the diode area and $x$ is given by the position of the edge of the depletion region:

$$x = \frac{\varepsilon_s S}{C}.\qquad (A1.13)$$

Profiling information can be obtained only for distances $x > x_0$, the zero bias depletion width. The most convenient way of profiling is the use of a capacitance meter or bridge to obtain the $C-V$ relationship and to use a suitable computer program to compute the $N(x)$ versus $x$ curve.

# Common Etches for Silicon

| Composition | | Comments |
|---|---|---|
| $HF:HNO_3:CH_3COOH$ | 3:5:3 | CP4—polishes the surface, required 2–3 min |
| $HF:HNO_3:COOH$ | 3:5:3 | CP8—polishes the surface |
| $HF:HNO_3$ | 1:3 | White etch, polishes the surface, requires $\simeq 15$ s |
| Conc. HF with 0.1–0.5% Conc. $HNO_3$ | | Delineates pn junctions one drop on freshly lapped surfaces |
| $H_2O:NH_2(CH_2)_2NH_2$ | | Does not attack silicon oxide |
| NaOH or KOH 1–30% solution | | Develops structural details |
| $HF:HNO_3:H_2O$   4 ml:2 ml: 4ml plus $AgNO_3$ (200 mg) | | Develops faults in epitaxial layers |
| Ethylenediamine catechol $H_2O$ (hydrazine) | | Orientation- and concentration-dependent. Stops etching at $p^{++}$ interface. Very slow etching of $SiO_2$. |
| KOH-normal propanol $H_2O$ | | Etches [100] a hundred times faster than [111]. Stops at $p^+$ interface |

# Common Etches for Gallium Arsenide

| Composition | | Comments |
| --- | --- | --- |
| $(1-20\%)Br_2$ in $CH_3COOH$ | | Etching rate drops below $5\%$ |
| $HF:HNO_3:H_2O$ | $1:3:2$ | Good for rapid etching, can be slowed down by increasing $H_2O$ |
| $H_2SO_4:H_2O_2:H_2O$ | $3:1:1$ | Suitable for mesa etching |

# The Mathematics of Diffusion and Ion Implantation

## A4.1 DIFFUSION

A simple model of diffusion is depicted in Fig. A4.1 where the atomic planes are separated by a distance $d/\sqrt{3}$ where $d$ is the spacing between the tetrahedral sites of the diamond (silicon and gallium arsenide) structure. The centre of each layer contains the centre of the lattice plane from which atoms may jump. Taking the example of the silicon lattice and defining the number of atoms in each plane (1) and (2) as $n_1$ and $n_2$ respectively, then the volume concentrations $N_1$ and $N_2$ are given by ($n_1$/volume) and ($n_2$/volume); hence:

$$N_1 = \frac{n_1\sqrt{3}}{Sd}, \qquad N_2 = \frac{n_2\sqrt{3}}{Sd}.$$

In this solid structure each atom has four nearest neighbouring sites it may jump into, two in the plane to the left and two in the plane to the right. We may therefore assume that half the atoms may move from plane (1) to the left and half to the right. Since the jump period is $(1/v_j)$ then the net flow of atoms from layer (1) and (2) is given by

$$\frac{\Delta n}{\Delta t} = \frac{(n_1 - n_2)/2}{(1/v_j)}. \tag{A4.1}$$

Substituting for $n_1$ and $n_2$

$$\frac{\Delta n}{\Delta t} = \frac{Sd \times v_j}{2\sqrt{3}}(N_1 - N_2), \tag{A4.2}$$

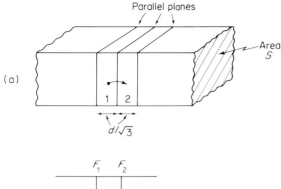

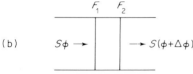

(a)

(b)    $S\phi \rightarrow$    $\rightarrow S(\phi + \Delta\phi)$

Fig. A4.1

but the concentration gradient $(\Delta N/\Delta x) = \sqrt{3}(N_2 - N_1)/d$, and hence

$$(N_1 - N_2) = -\frac{d}{\sqrt{3}}\left(\frac{\Delta N}{\Delta x}\right). \qquad (A4.3)$$

Substituting into (A4.2)

$$\frac{\Delta n}{\Delta t} = -\frac{Sd^2 v_j}{6}\left(\frac{\Delta N}{\Delta x}\right).$$

Since $(1/S)(\Delta n/\Delta t)$ is the net flux $\phi$ of diffusing atoms per unit area of solid, the flux

$$\phi = -\frac{d^2 v_j}{6}\left(\frac{\Delta N}{\Delta x}\right). \qquad (A4.4)$$

The constant $v_j d^2/6$ is defined as the diffusion constant $D$ and hence we write:

$$\boxed{\phi = -D\left(\frac{\partial N}{\partial x}\right).} \qquad (A4.5)$$

This equation is called *Fick's first law of diffusion.*

Consider now the diffusion coefficient $D = (d^2 v_j/6)$ in some more detail. Substituting for $v_j$ for a substitutional diffusion (Equation 3.11), $D$ may be written

$$D = \frac{2vd^2}{3}\exp\left\{\frac{-(E_a + E_s)}{kT}\right\} = D_0 \exp\left\{\frac{-(E_a + E_s)}{kT}\right\}. \qquad (A4.6)$$

The pre-exponential term $D_0$ contains the vibrational frequency $v$ and is hence a function of temperature. However, as a first approximation it may be assumed that $D_0$ is constant over a limited temperature range (i.e. $< 100$ K). Thus we may write:

$$\ln (D) = \ln (D_0) - \frac{(E_a + E_s)}{k} \left( \frac{1}{T} \right). \tag{A4.7}$$

A plot of the natural logarithm of $D$ versus $(1/T)$ will yield a straight line of slope $(E_a + E_s)/k$ whose extrapolated intercept to $(1/T) = 0$ gives the value of $D_0$. Such a plot is shown in Fig. A4.2 for aluminium, phosphorus, boron, and antimony in silicon.

A second useful approximation is to assume that $D$ does not depend upon concentration $N$, but this approximation is only valid for doping densities less than $10^{26}$ atoms m$^{-3}$.

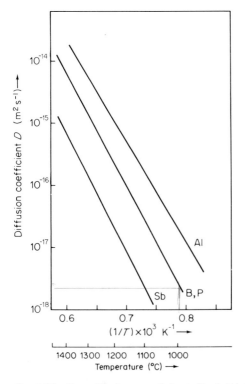

Fig. A4.2  The diffusion coefficient $D$ plotted as a function of $1/T$ for aluminium, phosphorus, boron, and antimony in silicon

Note that for interstitial diffusions the expression for $D$ becomes

$$D = D_0 \exp(-E_a/kT).$$

Figure A4.3 shows the appropriate plot of logarithm $(D)$ versus $1/T$ for two interstitial diffusants, gold and copper. In the comparison of the curves of Figs. A4.2 and A4.3, note the much lower slope of the interstitial diffusants and the much larger values of $D_0$.

Consider once again the simple one-dimensional model of diffusion shown in Fig. A4.1(b).

The flux entering face $F_1$ is $S\phi$ and the flux leaving at $F_2$ is $S(\phi + \Delta\phi)$; hence the net flux entering the volume $(S\,\Delta x)$ is just $-(S\,\Delta\phi)$, so we may write

$$\left(\frac{\partial N}{\partial t}\right) S\,\Delta x = -S\,\Delta\phi. \qquad (A4.8)$$

Substituting for $\Delta\phi/\Delta x$ from Fick's first law,

$$\frac{\partial\phi}{\partial x} = -\frac{\partial}{\partial x} D\left(\frac{\partial N}{\partial x}\right) \qquad (A4.9)$$

$$\left(\frac{\partial N}{\partial t}\right) = \frac{\partial}{\partial x} D\left(\frac{\partial N}{\partial x}\right) \qquad (A4.10)$$

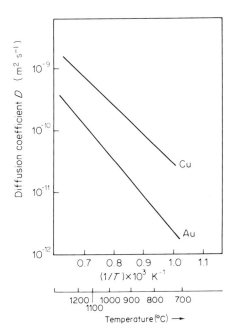

Fig. A4.3   The diffusion coefficient plotted as a function of $1/T$ for copper and gold in silicon

which, if $D$ is assumed to be independent of concentration, simplifies to *Fick's second law*:

$$\frac{\partial N}{\partial t} = D\frac{\partial^2 N}{\partial x^2}. \qquad (A4.11)$$

This differential equation is very important, as its solution, subject to specific boundary conditions, enables controlled diffusion profiles to be obtained and hence pn junctions to be fabricated. In most practical conditions two boundary conditions allow relatively simple solutions to Fick's law to be used.[†] These were discussed in Chapter 3.

## A4.2   ION IMPLANTATION

The basic theoretical equations which are used to predict the range distribution of ions implanted into solids is that developed by the Danish group at the university of Aarhus in Denmark (Lindhard Scharff and Schiott—the LSS theory). In this theory the following assumptions are made:

(a)  The solid is assumed to be amorphous with no open channels to allow the implanted ions to become trapped and hence to have anomalous deep ranges (i.e. channelling).

(b)  The incident ion beam is mono-energetic and does not contain a mixture of isotopes.

In real single crystals, the phenomenon of 'channelling' occurs. In this case ions become trapped in the open planes and axes and can have ranges of many times the normal. Whilst it is hard to completely inhibit this effect it can be minimized by misorientating the incident beam with respect to the open axial and planar channels.

The LSS theory predicts that the range distribution $N(x)$ will be approximately Gaussian (Fig. 3.16) with a projected range $\bar{R}_p$, where

$$N(x) = \frac{\phi}{\sqrt{(2\pi)}\sigma_p}\exp\left[-\frac{(x-\bar{R}_p)^2}{2\sigma_p^2}\right], \qquad (A4.12)$$

where $\phi$ is the total number of ions implanted (fluence, $m^{-3}$) and $\sigma_p$ the standard deviation in the projected range.

Note that at $x = \bar{R}_p$ the peak concentration $N(x) = N_p$ becomes

$$N_p = \phi/(\sqrt{2\pi})\sigma_p \simeq 0.4\phi/\sigma_p.$$

---

[†] For the reader who wishes to pursue this topic in more detail refer to Ref. 2 (Appendix A.6).

The total number of ions implanted is

$$\phi = \int_{x=0}^{\infty} N(x)\mathrm{d}\,x,$$

that is, the area under the implanted profile curve.

A comprehensive outline of the theory of Lindhard and co-workers is given by Gibbons *et al.*, who also list the numerical calculations for a range of ions in silicon and gallium arsenide. Some of this data is plotted in Figs. A4.4(a) and (b).

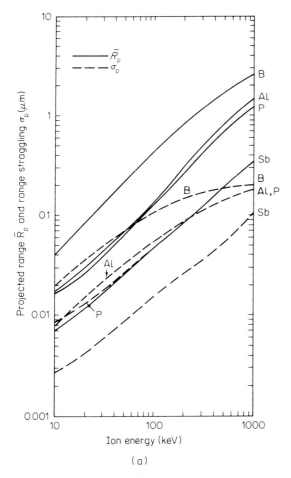

(a)

Fig. A4.4   Range data $\bar{R}_p$ and $\sigma_p$ for some p- and n-type dopants in (a) silicon and (b) (*see overleaf*) gallium arsenide

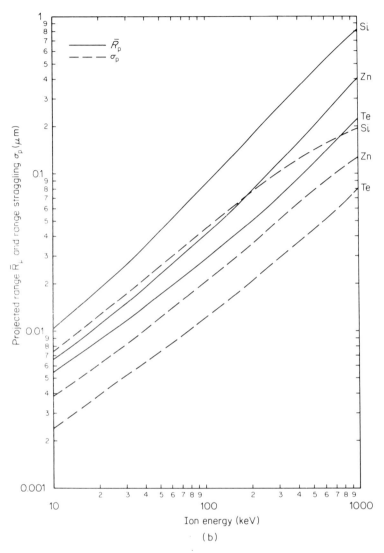

Fig. A4.4   (*continued*)

## References

J. Lindhard, M Scharff, and H. E. Schiott, *Mat. Fys. Medd Vid. Selek*, **33**, 14 (1963).
J. F. Gibbons, W. S. Johnson, and S. W. Mylroie, *Projected Range Statistics: Semiconductors and Related Material* (Dowden, Hutchinson and Ross, 1975).

# Thermal Oxidation of Silicon

Consider a unit-area cross section of semiconductor plus oxide, in which the oxidant concentration $N$ (molecules/m$^3$) varies linearly from $N_s$ at the surface to $N_i$ at the interface, as shown in fig. A5.1. If the flux $F_o$ of oxidant molecules passing through the oxide satisfies Fick's first law of diffusion (see Appendix 4),

$$F_o = -D_o \frac{dN}{dx} = \frac{D_o(N_s - N_i)}{x_o}. \qquad (A5.1)$$

The flux $F_i$ represents the rate at which the oxidant molecules are consumed at the interface in the production of SiO$_2$. It is reasonable to assume that $F_i \propto N_i$:

$$F_i = k_i N_i. \qquad (A5.2)$$

In the steady state the two fluxes must be equal ($F_o = F_i = F_f$, say). Equating (A5.1) and (A5.2),

$$N_i = \frac{N_s}{1 + k_i x_o/D_o}$$

and

$$F_f = \frac{N_s}{1/k_i + x_o/D_o}.$$

In $N_m$ oxidant molecules are required to produce unit volume of SiO$_2$, then the rate at which the SiO$_2$ film thickness increases is given by

$$\frac{dx_o}{dt} = \frac{F_f}{N_m} = \frac{N_s/N_m}{1/k_i + x_o/D_o}.$$

193

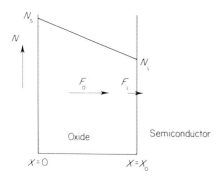

Fig. A5.1

Integrating from 0 to $t$ with $x_o|_{t=0} = x_i$ gives

$$x_o - x_i + \frac{k_i}{2D_o}(x_0^2 - x_i^2) = \frac{N_s k_i}{N_m}. \tag{A5.3}$$

Equation (A5.3) is a quadratic in $x_o$ and may be solved to give

$$x_o = \frac{A}{2}\left[\sqrt{\left(1 + \frac{t + \tau_o}{A^2/4B}\right)} - 1\right] \tag{A5.4}$$

where $A = 2D_o/k_i$, $B = 2D_o N_s/N_m$, $\tau_o = (x_i^2 + Ax_i)/B$.

# List of Extra Reading

| *Reference* | *Comment* |
|---|---|
| 1. H. M. Olsen, Chapter 2 (*Device technology*) in *Variable Impedance Devices*, M. J. Howes and D. V. Morgan (eds.) Wiley (1978) | An excellent chapter dealing with pn diodes and Schottky diodes. Good on GaAs. |
| 2. S. K. Ghandi, *Theory and Practice of Microelectronics*,Wiley (1968) | A comprehensive text covering technology and device design. Excellent coverage, good reference book. A new edition brings the book up to date. |
| 3. R. A. Colclaser, *Microelectronics*, Wiley (1980) | Very good text. Wide coverage and up to date. |
| 4. P. E. Gise and R. Blanchard, *Semiconductor and Integrated Circuits Fabrication Techniques*, Prentice-Hall (1979) | A good laboratory style book—very useful to the beginner. |
| 5. P. Daniels, Chapter 5 in *Large Scale Integration*, M. J. Howes and D. V. Morgan (eds.), Wiley (1981) | A concise and up-to-date survey of technology. |
| 6. S. M. Sze, *The Physics of Semiconductor Devices*, 2nd Edition, Wiley (1981) | One of the most comprehensive semiconductor books and an excellent reference book. |
| 7. E. H. Rhoderick, *Metal Semiconductor Contacts*, Oxford University Press (1958) | A comprehensive and up-to-date text on this topic. To be recommended to the advanced student. A second edition published in (1988) bring the book up to date (Rhoderick & Williams) |

8. G. Carter and W. A. Grant, *Ion Implantation in Semiconductors*, E. J. Arnold (1976)

    A good introductory book.

9. A. Grove, *Physics and Technology of Semiconductor Devices*, Wiley (1967)

    Excellent treatment of technology although now dated.

10. T. P. Kaberservice, *Applied Mircoelectronics*, West (1978)

    Good section on electronic factors in LSI.

11. E. S. Yang, *Fundamentals of Semiconductor Devices*, McGraw-Hill (1978)

    Excellent up-to-date text of device theory and technology.

12. C. Mead and L. Conway, *Introduction to VLSI Systems*, Addison-Wesley (1980)

    This has become the standard text on chip level design.

13. J. Watson, *Semiconductor Circuit Design*, 4th Edition, Adam-Hilger (1983)

    Broad based introductory text which includes the transition to circuit design.

# 7
APPENDIX

# Additional Exercises

1.  Write down a general expression for the conductivity of slab of a uniform semiconductor containing both electrons and holes.

    How does this expression simplify for a sample of:

    (a)  intrinsic material, and
    (b)  extrinsic n-type material?

2.  Figure A7.1 shows a plot of the logarithm of conductivity versus $1/T$ for a semiconductor—what can you deduce from this?

3.  A bar of silicon contains $2 \times 10^{24}$ boron atoms m$^{-3}$ and $3.5 \times 10^{24}$ antimony atoms m$^{-3}$. Using Fig. 3.7 determine the mobilities of the electrons and holes ($\mu_n$ and $\mu_p$) and the resistivity of the bar.

    Why does the resistivity of the bar differ from one containing just $1.5 \times 10^{24}$ antimony atoms m$^{-3}$? The data in Fig. 3.7 may be used for your calculations.

4.  Using a suitable figure define what is meant by the sheet resistivity.

5.  Figure A7.2 shows the structure of a commercial GaAs transferred electron device and also a plot of the device current ratio $J_{B=0}/J_B$ versus the square of the magnetic field $B^2$ for a device biased below threshold (directions as shown in figure). Show with the appropriate equations what you may deduce from this.

6.  Define the term 'relaxation time ($\tau$)' for a charge carrier moving in a semiconductor and discuss factors which control $\tau$ in real semiconductors.

7.  Given a slice of semiconductor material in the form shown in Fig. A7.3(a) and the experimental information shown in Fig. A7.3(b), calculate the conductivity, doping density, nature of semiconductor (p- or n-type), and mobility. (Derive all equations used.)

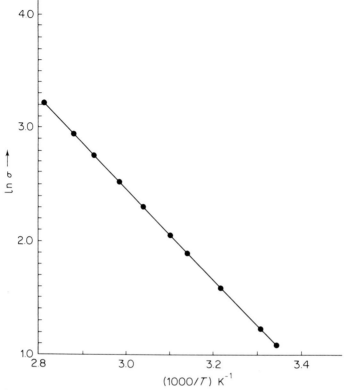

Fig. A7.1

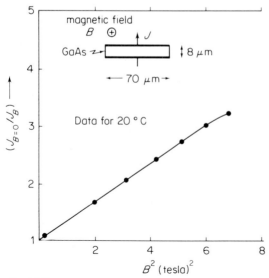

Fig. A7.2

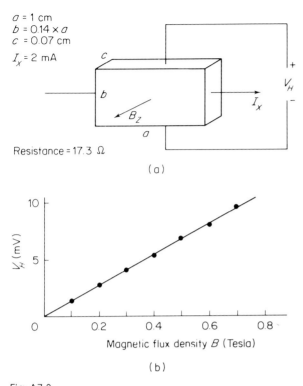

$a$ = 1 cm
$b$ = 0.14 × $a$
$c$ = 0.07 cm

$I_x$ = 2 mA

Resistance = 17.3 Ω

(a)

Magnetic flux density $B$ (Tesla)

(b)

Fig. A7.3

8.  In an experiment on a semiconducting specimen of resistivity 0.05 Ω m, the Hall coefficient was found to be $45 \times 10^{-3}$ m$^3$ C$^{-1}$. What are the free charge carriers, what is their concentration, and what is their mobility?

9.  The electrical conductivity of a piece of intrinsic material doubles when its temperature is raised from 300 K to 310 K; discuss fully how you could use this information to calculate the band gap $E_g$ of the material (discuss any equations and approximations used in your calculations). (*Note*: $kT = 0.025$ eV at 300 K.)

10. The equilibrium current through a slice of n-type germanium is increased by 0.1 % when it is illuminated by a light which forms hole–electron pairs at the rate of $2 \times 10^{22}$ m$^{-3}$ s$^{-1}$. Calculate the lifetime (resistivity = 1.74 Ω m and mobility $\mu = 1.74 \times 10^{-1}$ m$^2$ V$^{-1}$ s$^{-1}$).

11. Discuss the basic assumptions which lead to the conclusion that one-dimensional diffusion processes in a semiconductor may be described by:

$$\frac{\partial N}{\partial t} = D \frac{\partial^2 N}{\partial x^2}$$

where $N$ is the chemical concentration of the diffusing atoms.

12. Describe the physical mechanism by which diffusion of dopant atoms may occur in silicon (restrict your discussion to the so-called substitutional dopants).

    Diffusion is governed by Fick's second law:

    $$\frac{\partial N}{\partial t} = D \frac{\partial^2 N}{\partial x^2}.$$

    Discuss briefly the solutions to this equation for (a) constant source diffusion and (b) limited source diffusion. Illustrate schematically how the diffusion profiles so obtained will vary with temperature and time. (*Hint*: The diffusion coefficient $D = D_0 \exp\left(-(E_a + E_s)/kT\right)$.)

13. A p-type silicon sample has a uniform doping density of $N_A = 10^{23}$ atoms $m^{-3}$. Determine the diode junction depth when the sample is subjected to the following process: Predeposition with excess phosphorus at 950 °C for 30 minutes (i.e. note that the solid solubility concentration of phosphorus at 950 °C is $8 \times 10^{26}$ atoms $m^{-3}$. Plot the resulting diffusion profile.

    Figure A4.2 is a plot of $D$ versus $(1000/T)$ for a range of atoms in silicon. The table is a listing of $Z$ versus erfc $(Z)$

    | $Z$ | 0 | 0.094 | 1.808 | 2.713 | 3.617 | 4.520 |
    |---|---|---|---|---|---|---|
    | erfc $(Z)$ | 1 | 0.2 | 0.01 | $1.23 \times 10^{-4}$ | $3.1 \times 10^{-7}$ | $1.6 \times 10^{-10}$ |

14. A donor impurity is diffused into a semiconductor which has a non-diffusing background acceptor impurity (concentration $10^{21}$ $m^{-3}$), from a constant source (concentration $10^{23}$ $m^{-3}$). Given that the diffusion time is 10 hours and the diffusion temperature 1100 °C, plot the resultant doping profile through the semiconductor. You may assume that the activation energy of the diffusion process is 3 eV and the apparent value of the diffusion coefficient (of the diffusing atoms) at $T = \infty$ is $10^{-2}$ $m^2$ $s^{-1}$.

15. What are the advantages of ion implantation over diffusion for device fabrication?

16. Describe why the occurrence of a natural oxide ($SiO_2$), with its excellent electrical and physical properties, was so important to the development of silicon planar technology and hence silicon integrated circuits.

17. Discuss in detail the process of thermal oxidation of silicon, indicating the basic mechanism involved in the process.

18. By assuming that the surface concentration of oxygen is governed by its solid solubility limit $N_s$ and using a linear approximation for the concen-

tration profile, show that the relationship between oxide thickness $x$ and time $t$ is given by the relationship

$$x = \frac{D}{k_i}\left[\left(1 + \frac{2N_s k_i^2 t}{DN_m}\right)^{1/2} - 1\right],$$

where $D$ is effective diffusion constant for the oxygen in the oxide, $N_m$ the number of molecules of oxygen per unit volume of the oxide and $k_i$ the reaction rate constant.

19. What are the advantages and disadvantages in using an in-contact mask to form patterns in photoresist?

20. Explain why positive resists produce images rather larger, and negative resists smaller than the mask dimensions.

21. Given that a surface alignment mark has several microns of epitaxial silicon grown over it, explain how the next masking level can be aligned to it.

22. Design a mask set for two junction diodes connected back to back.

23. Describe the advantages and limitations of masking to higher levels of integration the design of electronic systems.

24. Discuss the relative merits of wet and dry thermal oxidation.

25. Compare the times taken to grow a 1 $\mu$m silicon oxide film at 1100 °C using

(a)  a dry oxide,
(b)  a steam-grown oxide.

26. An oxide film is grown for 1 hour at 950 °C in (a) dry oxygen, (b) steam. Calculate the resulting thicknesses and comment.

27. What are the limits to the optical photolithography process when fabricating sub-micron device structures? Describe in detail how these limitations may be overcome.

28. Describe how the process of photolithography is used in micro-technology and using the example of a positive resist show with clear diagrams how the process may be used to fabricate mesa diodes on an oxide-coated silicon wafer.

29. Starting with a uniformly doped silicon epitaxial layer ($N_D \simeq 10^{23}$ m$^{-3}$) 10 $\mu$m thick grown on an n$^+$-substrate ($N_D > 10^{24}$ m$^{-3}$) and using the data in Fig. A4.4, determine a scheme to produce a p$^+$n junction where the metallurgical junction is 0.2 $\mu$m below the surface. Discuss the reasons for your choice of p-type dopant and the procedure needed to activate the implanted ions.

30. Outline the detailed technological steps and sketch the masking schemes needed to fabricate a MESFET transistor suitable for operation up to 10 GHz starting from:

    (a) an n-type epitaxial layer on a semi-insulating substrate,
    (b) a uniform semi-insulating substrate. Start by specifying your initial material requirement. (*Hint*: Study Fig. 7.22.)

31. Compare the relative merit of the two diode structures illustrated in Figs. 7.20(b) and (c).

# 8

# Properties of Si, GaAs and SiO$_2$

| Property | Si | GaAs | SiO$_2$ |
|---|---|---|---|
| Atoms or molecules m$^{-3}$ | $5.0 \times 10^{28}$ | $4.42 \times 10^{28}$ | $2.3 \times 10^{28}$ |
| Atomic or molecular weight | 28.08 | 144.63 | 60.08 |
| Density (g m$^{-3}$) | $2.33 \times 10^6$ | $5.32 \times 10^6$ | $2.27 \times 10^6$ |
| Breakdown field (V m$^{-1}$) | $\sim 3 \times 10^7$ | $\sim 3.5 \times 10^7$ | $\sim 6 \times 10^8$ |
| Crystal structure | Diamond | Zinc blende | Amorphous |
| Dielectric constant | 11.9 | 13.1 | 3.9 |
| Electron affinity $\chi$(V) | 4.05 | 4.07 | 0.9 |
| Energy gap (eV) | 1.12 | 1.42 | $\sim 8$ |
| Intrinsic carrier concentration $n_i$ (m$^{-3}$) | $1.45 \times 10^{16}$ | $1.79 \times 10^{12}$ | |
| Lattice constant (Å) | 5.431 | 5.654 | |
| Effective mass: | | | |
| Electrons | $m_e = 0.33m, m_e^* = 0.26m$ | 0.668m | |
| Holes | $m_h = 0.56m, m_k^* = 0.38m$ | 0.56m | |
| Intrinsic mobility: | | | |
| Electron (m$^2$ V$^{-1}$ s$^{-1}$) | 0.15 | 0.8500 | |
| Hole (m$^2$ V$^{-1}$ s$^{-1}$) | 0.045 | 0.04 | |
| Temperature coefficient of expansion | $2.5 \times 10^{-6}$ | $5.8 \times 10^{-6}$ | $5 \times 10^{-7}$ |
| Thermal conductivity (W m$^{-1}$ deg C$^{-1}$) | 150 | 46.0 | 1 |

# Subject Index